IET PROFESSIONAL APPLICATIONS OF COMPUTING SERIES 14

# Foundations for Model-based Systems Engineering

**Other volumes in this series:**

# Foundations for Model-based Systems Engineering

## From Patterns to Models

Jon Holt, Simon Perry
and Mike Brownsword

The Institution of Engineering and Technology

Published by The Institution of Engineering and Technology, London, United Kingdom

The Institution of Engineering and Technology is registered as a Charity in England & Wales (no. 211014) and Scotland (no. SC038698).

First published 2016

The Institution of Engineering and Technology
Michael Faraday House
Six Hills Way, Stevenage
Herts, SG1 2AY, United Kingdom

www.theiet.org

**British Library Cataloguing in Publication Data**
A catalogue record for this product is available from the British Library


**ISBN 978-1-78561-050-9 (hardback)**
**ISBN 978-1-75861-051-6 (PDF)**



Typeset in India by MPS Limited
Printed in the UK by CPI Group (UK) Ltd, Croydon

*This book is dedicated to the memory of Stuart Arnold – a
brilliant Systems Engineer and a very fine man.*

*–Jon Holt*

*To Myrtle Warner Shingler, 1924–2014.*

*–Simon Perry*

*To the memory of Robert Arthur Johnson (1937–2011)
who always saw things in a positive light,
whatever the situation.*

*–Mike Brownsword*

# Contents

# Acknowledgements

As usual, this book would not have been possible without the many colleagues, friends and clients who have let us loose on their Systems over the last 20-plus years.

This book is a departure from our usual offering in that, due to new guidelines, we are no longer allowed our erudite and informative quotes at the beginning of each chapter. Witty quotes, we mourn you.

My now-predictable thanks go to, as ever, Mike and Sue Rodd who are still major influences on my life. Also, all my love and thanks go to my wonderful wife Rebecca, my sometimes-wonderful children: Jude, Eliza and Roo and finally to Olive and Houdini who provide feline relief in my life.

Jon Holt, April 2016


The ideas in this book are a result of thoughts sparked by conversations with many clients over many years. Mentioning them all by name would require a list as big as the current book. However, some must be so mentioned as having particular input into our latest volume: thanks to Colin Wood (and the INCOSE UK MBSE WG) for sparking off thoughts that lead to the Evidence Pattern, thanks to Clive Ingleby for back-fitting a meaning for PaDRe and to Ian, Manoj and Andy for letting us apply our thoughts to your system.

As Jon has said, sadly this is the first of our books without chapter quotes. However, they are available, if you know where to look.

As always, this book would not have been possible without the love, friendship and encouragement of my beautiful wife Sally. Every day I am thankful for the day I met her and for the day she said "yes." Finally, Walter and Dolly are still with us, demonstrating daily what effective bed-warmers two cats make!

Simon Perry, April 2016


Not one to miss a pattern; I would like to recognise all those who have influenced the thinking in this book including the Systems and Safety team at Atkins, Colin and the MBSE Working Group and all those current and past at the Digital Railway programme. DR has enabled many of these ideas to be applied, so specific thanks go to those who have been instrumental in bringing Architecture to the centre of the programme and using it for all aspects of system development and analysis.

Although I really enjoy the quotes I often find, as I am not as witty as Jon and Simon, they take longer to write than the book, so I will mourn them less.

Finally, to my wonderful wife who keeps me on the straight and narrow, Sam and Daniel for the physical and mental exercise they provide and our current herd of cats Ben, Shadow, Kai and Jay who have contributed more typos and additional text than I care to remember.

Mike Brownsword, April 2016

*Part I*

**Introduction and approach**

*Chapter 1*

# Introduction

## 1.1 Introduction

The world of Systems Engineering is changing. As the whole discipline matures, so the use of model-based techniques and approaches is becoming more and more prevalent. Indeed, the International Council on Systems Engineering (INCOSE) predicts that by 2025 all Systems Engineering will be model based:

> *'Model-based Systems Engineering will become the "norm" for systems engineering execution, with specific focus placed on integrated modeling (sic) environments. These systems models go "beyond the boxes", incorporating geometric, production and operational views. Integrated models reduce inconsistencies, enable automation and support early and continual verification by analysis'.* (INCOSE, 2014)

The question now is not 'should we model?' but, rather, 'how do we model effectively and efficiently?' As the need for rigorous and robust techniques increases so the use of MBSE structures, such as ontologies, frameworks and processes, becomes ever more important.

This book discusses the use of MBSE Patterns that can be used to enhance and augment existing MBSE approaches.

This chapter discusses existing MBSE practice and examines the role of Patterns within an MBSE approach.

## 1.2 Introducing Patterns

The use of Patterns is an acknowledged and established part of other engineering disciplines, in particular in the world of software engineering.

The definition of Pattern that will be used throughout this book is

> *'A Pattern is a defined set of Viewpoints and associated Views that may be re-used for a number of different applications'*

The term 'Viewpoints' here refers to a set of templates for the Pattern that form the basis for the 'Views'. The Views are the end product of the Pattern that are generated as part of the system development activities. A View is a realisation of a Viewpoint.

The main aspect of the use of Patterns is to allow re-use which brings with it a number of tangible benefits:

- Shortened development time. By working to a set of pre-defined Viewpoints the structure and content of each View has already been decided, hence reducing the amount of time required to generate the individual Views.
- Improved consistency. By working in accordance with the Pattern Viewpoints (and, hence, Ontology) then consistency between the resultant Pattern Views is guaranteed.
- Shortened learning curve. If systems modellers are educated and trained in the use of Patterns then they can apply this knowledge in a variety of applications. This more-consistent use will result in a shortened learning curve, rather than learning new techniques each time a new application is modelled.
- Increased efficiency through tool implementation of Patterns. The Viewpoints that describe the Pattern may be implemented in a modelling tool through use of profiles. These profiles guide the modeller through which Views need to be created and their content, along with a set of rules that may be used as a basis for automated consistency checks.

A more-detailed discussion of the history of Patterns may be found in Chapter 3.

## 1.3    Realising MBSE – People, Process and Tools

In order to realise MBSE effectively and efficiently, there are three aspects that need to be considered, as shown in Figure 1.1.

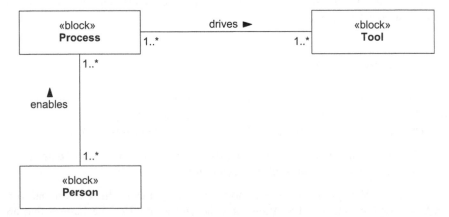

*Figure 1.1    Realising MBSE*

Figure 1.1 shows the three essential aspects of MBSE that need to be considered for successful MBSE, which are:

- People (represented by one or more 'Person'), by which we mean competent people with the appropriate skills necessary to perform their roles successfully. This means understanding the Stakeholder Roles that are necessary for enabling the 'Process' and understanding the competencies required in the form of a Competency Scope for each – see Holt and Perry (2013, chapter 8).
- 'Process', by which we mean the fundamental approach that is being followed for MBSE. This means having well-defined and accepted Ontologies, Frameworks and Processes in place. The 'Process' is enabled by one or more 'Person' and drives the implementation using one or more 'Tool'.
- 'Tool', by which we mean the use of any CASE tool, application (whether manual or automated), notation or technique required for MBSE. The role of the Tool is to implement the Process that is driving it.

The use of MBSE Patterns can enhance and augment all three of these aspects as the benefits identified in the previous section apply to these three aspects, for example:

- People. The MBSE Patterns in this book can be used as part of the Competency assessment Process that is used to demonstrate the Competence of each Person. The second way that Patterns may enhance the People aspect is to shorten the learning curve associated with MBSE. If people are educated and trained in the use of Patterns then this knowledge may be put into play whenever each Pattern is re-used.
- Process. The MBSE Patterns form an integral part of defining Ontologies, Frameworks and Processes. For example, Frameworks may be constructed by re-using and tailoring established Patterns. If each Pattern has its associated Viewpoints defined then the definition of the Framework becomes so much quicker and easier as the structure and contents of each View will already be defined. See Chapter 21 for a discussion of this.
- Tools. Good MBSE modelling tools may be tailored to support a defined MBSE approach by defining the so-called "profiles". The use of profiles allows the tool to be tailored to implement a specific approach to MBSE. As part of this tailoring, it is possible to embed the Patterns into the tool, hence guiding their use and making the Pattern knowledge available to users of the tool.

The use of Ontologies, Frameworks and Processed will be discussed in more detail in Chapter 2.

## 1.4 Origins of the Patterns

The Patterns in this book have been generated over the last 15–20 years based on real-life experiences on real projects.

The use of MBSE has been increasing rapidly over the last decade or so and, in particular in the last five years. The more that MBSE is applied then the more modellers find themselves using the same constructs time and time again on different applications. This has resulted in many modellers adopting their own common approach to how different aspects of a System may be modelled.

It is untrue and unwise to assume that if something works for one project then it will work on another project and modelling is no different. However, if something works well on three or four projects then there is an increased likelihood that it will work on more projects.

When this scenario occurs, where some aspect of modelling has been applied successfully to a number of projects, then there is a good chance that there is a Pattern somewhere in the model that may be exploited for use elsewhere.

All of the Patterns on this book have been abstracted and defined based on their use in multiple projects, some of which have been in use by the authors for over 15 years.

## 1.5    How to use this book

### 1.5.1    Target audience

This book is targeted towards MBSE practitioners rather than newcomers to MBSE. That is not to say that this book cannot add value to newcomers, but the basics of MBSE along with the SysML notation are not taught in this book. This book has been written as a companion volume to *SysML for Systems Engineering – A Model-Based Approach* and the two books should be used together.

This book is intended to be used as a working reference, rather than a teaching aid.

### 1.5.2    Writing conventions adopted in the book

When thinking about MBSE and describing and discussing the different aspects of MBSE, there is a lot of potential for confusion. Some terms that are used in the SysML notation, for example, are also widely accepted MBSE terms and are also everyday words. Therefore, when using a specific term, we want to differentiate between the MBSE term, the SysML term, specific elements from diagrams and everyday usage of such terms. In order to minimise this confusion, the following writing conventions are adopted in this book.

- All terms from the SysML notation that form part of the standard, are written in italics. Therefore, the use of *block* refers to the SysML construct, whereas the same word without italics – block – refers to an impediment (or piece of cheese).
- All terms that are defined as part of the overall model presented in this book, such as the MBSE Ontology, MBSE Framework, are presented with capitalised words. Therefore the use of Project refers to the Ontology Element, whereas the same word without capitals – project – refers to a non-specific usage of the term as a noun or verb.
- All words that are being referenced from a specific diagram are shown in quotes. Therefore, the use of 'Ontology Element' is referring to a specific element in a specific diagram.

● All View names are shown as singular. Therefore, the term Process Behaviour View may be referring to any number of diagrams, rather than a single one.
● Any word that requires emphasis is shown in "double quotes."

Some examples of this are

*Table 1.1    Example sentencing to illustrate the convention*

| Example sentence | Meaning |
|---|---|
| A Use Case may be visualised as a *use case* | Use Case – term from the MBSE Ontology<br>*use case* – the term from SysML notation |
| Engineering activity can be shown as an Activity on a Process and may be visualised as an *activity* | activity – the everyday usage of the word<br>Activity – the term from the MBSE Ontology<br>*activity* – the term from the SysML notation |
| The diagram here shows the 'MBSE Process Framework' is made up of one or more 'Process Behaviour View' | 'MBSE Process Framework' – a specific term from a specific diagram that is being described<br>'Process Behaviour View' – a specific term from a specific diagram that is being described |
| When defining Processes it is typical to create a number of *activity diagrams* that will visualise the Process Behaviour View | Processes – the term from the MBSE Ontology<br>*activity diagram* – the term from SysML notation<br>Process Behaviour View – the term from the MBSE Framework |
| It is important to understand the "why" of MBSE | "why" – emphasis of the word |

Table 1.1 here shows some example sentences, the convention adopted and how they should be read. In summary, look out for 'Quotes', Capitalisation and *italics* as these have specific meaning. Finally, remember that the use of "double quotes" simple represents emphasis.

In terms of the diagrams that are presented in this book, and there are very many, they are used for two different purposes and may be easily differentiated between:

● In general, diagrams used by way of an explanation, such as all of those found in this chapter, are presented without a *frame*. The exception is for those diagrams that explicitly represent Views based on the Framework for Architectural Frameworks (FAF; introduced in Chapter 3). In this case, these diagrams will have *frames* around them in order to emphasise that they are a formal and explicit part of a FAF definition. This typically applies to all the Patterns definitions in Part II. A *frame* is represented visually by a named box around the border of the diagram.
● Diagrams used as specific examples, such as those showing examples of the use of Patterns in Part II and examples found in Part IV that presents a case study, will each have a *frame* around them.

By adopting this writing convention, we "hope" to avoid confusion later in the book.

### *1.5.3   Book overview*

This book is structured as shown in Figure 1.2.

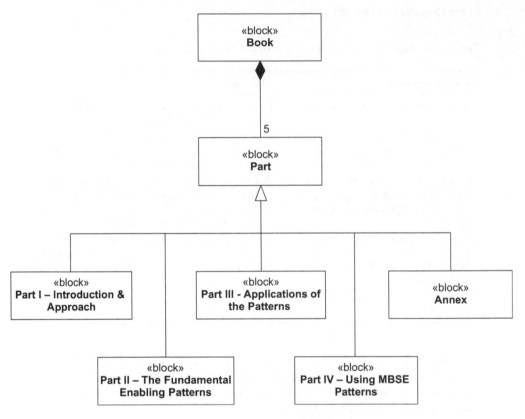

*Figure 1.2   Structure of the book*

Figure 1.2 shows that the 'Book' is composed of five 'Part(s)' which are:

- 'Part I – Introduction and Approach'. This part comprises three chapters that provide an introduction (this chapter), an overview of the general approach taken in this book (Chapter 2) and an in-depth introduction to the world of Patterns (Chapter 3).
- 'Part II – The Fundamental Enabling Patterns'. This part comprises ten chapters each of which introduces and discusses a single MBSE Pattern.
- 'Part III – Applications of the Patterns'. This part comprises five chapters each of which considers how the MBSE Patterns may be used to construct a number of Frameworks.
- 'Part IV – Using the MBSE Patterns'. This part comprises four Chapters each of which discusses how the MBSE Patterns may be used for diverse areas such

as defining bespoke Patterns, using Patterns for model assessment, using Patterns for model definition and retro-fitting Patterns to existing models.

- 'Annex' – This part comprises four Appendices that cover a summary of the SysML notation, a summary of the Framework for Architectural Frameworks (the FAF), the MBSE Ontology and an example Process for Pattern development.

References and further reading are provided at the end of each Chapter.

## References

INCOSE *A World in Motion – Systems Engineering Vision 2025,* INCOSE Publishing 2014, available at: http://www.incose.org/docs/default-source/aboutse/se-vision-2025.pdf?sfvrsn=4 [Accessed: October 2015].

Holt, J. and Perry, S. *SysML for Systems Engineering; 2nd Edition: A Model-based Approach.* London: IET Publishing; 2013.

# Chapter 2
# Approach

## 2.1  Introduction

Systems Engineering defines an approach for realising successful Systems (INCOSE, 2014). Model-based Systems Engineering provides an approach for using Models to realise the artefacts of a System based on a set of modelling Views (Holt and Perry, 2013).

This section introduces the main concepts that comprise the MBSE Meta-model that will be used throughout this book.

The first step is to define exactly what we mean when we refer to 'Systems' and 'Models'.

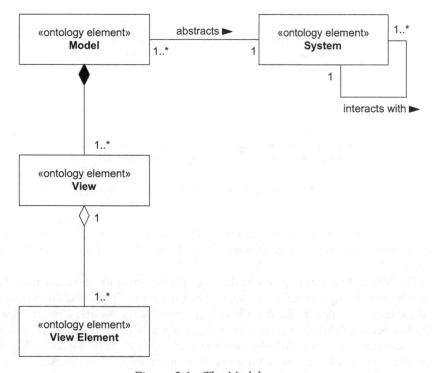

*Figure 2.1   The Model concept*

Figure 2.1 shows the fundamental concept of a 'Model' that abstracts a 'System'. This is crucial to everything in this book. The Model may represent any aspect of the System from any point of view.

The fundamental building block of the Model is the concept of a 'View' – one or more Views make up a Model. These Views may be visualised using any notation, text, tables, formal techniques and, in fact, in any way that is appropriate. In the context of this book, however, each View will be visualised, in general, using SysML and therefore each View will be manifested in one or more SysML diagram.

Each View is, in turn, made up of one or more 'View Element'. The View Element represents an instantiation of an Ontology Element in the Model and each will be visualised by one or more SysML element.

The Model, however, is not just a random collection of Views in the form of SysML diagrams, but has a rigorous structure. Unfortunately, it is all-too-often the case that the Model is merely a collection of pictures, rather than a true model. The structure of the Model is defined by a Framework, as shown in Figure 2.2.

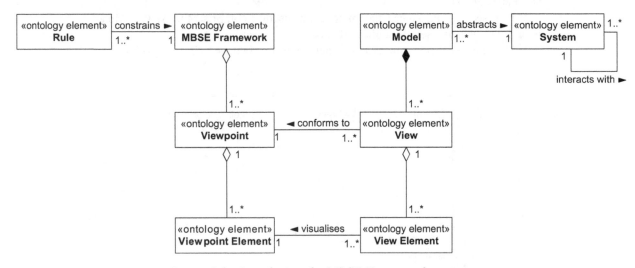

*Figure 2.2   Introducing the MBSE Framework concept*

Figure 2.2 introduces the idea of the 'MBSE Framework', which is made up of one or more 'Viewpoint' each of which is made up of one or more 'Viewpoint Element'.

The MBSE Framework provides the basis for the structure and content of the Model by identifying and defining a number of Viewpoints. These Viewpoints may be thought of as templates for the Views that comprise the Model. Once a Viewpoint has been defined, then each View must conform to its associated Viewpoint.

In the same way, the View Elements also conform to their associated Viewpoint Element. In this way, each View and all of its constituent elements conforms to the overall MBSE Framework.

In order to define a rigorous and consistent MBSE Framework, it is essential that an Ontology is in place, as introduced in Figure 2.3.

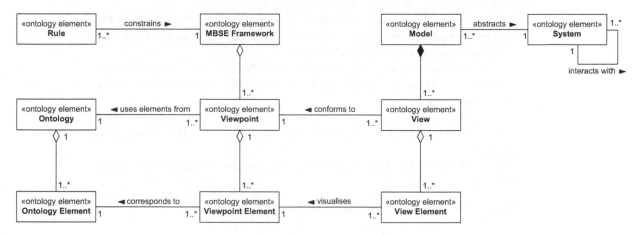

*Figure 2.3    Introducing the Ontology*

Figure 2.3 introduces the 'Ontology' that is made up of one or more 'Ontology Element'.

The Viewpoints use the Ontology Elements in the Ontology in order to define the contents of each Viewpoint. Each View, therefore, is essentially a collection of instances of Ontology Elements arranged according to a Viewpoint.

It is the Ontology that provides the structure, consistency and hence rigour for the MBSE Framework. The Ontology is discussed in more detail in the following section.

## 2.2    The MBSE Ontology

The MBSE Ontology forms the heart of the MBSE endeavour. It is the approach that is both advocated by this book and that has also been used to drive all of the content of this book. In this book the so-called MBSE Ontology is used for the following activities:

- *Defining concepts and terms*. The MBSE Ontology provides a visualisation of all the key concepts, the terminology used to describe them and the inter-relationships between said concepts. The MBSE Ontology, however, plays a pivotal role in the definition and use of any rigorous MBSE Framework and Patterns.
- *Defining Frameworks and Patterns that can be used for different aspects of MBSE*. Examples of these Frameworks in the companion volume to this book include: Processes, Needs, Architectures, etc. Whenever any Framework or Pattern is defined in terms of a set of Viewpoints, then an Ontology is essential. It is the MBSE Ontology that enforces the consistency and rigour demanded by the Patterns detailed in this book.

- *Defining MBSE Competencies*. When defining Competencies for MBSE activities, there needs to be a core knowledge base that underpins the Competencies, and this is realised by the MBSE Ontology.
- *Course content generation*. The MBSE Ontology may be thought of as the heart of the body of knowledge for MBSE and, therefore, is an ideal basis for any course content that must be developed.
- *Process implementation*. Many aspects of the Process, such as the Process model, the Artefacts, the execution in Life Cycles, will be based directly on the MBSE Ontology.
- *Tool implementation*. In order to optimise the use and therefore benefits of tool implementation of MBSE, then a full understanding of the MBSE Ontology is essential.
- *Defining Patterns*. All of the work presented in this book is based on the MBSE Ontology.

The structure of the MBSE Ontology is introduced in Figure 2.4.

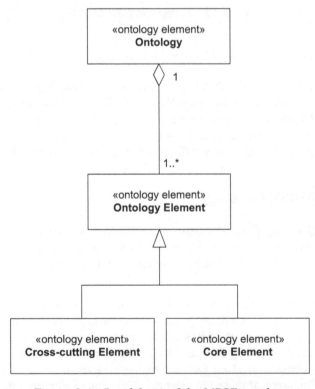

*Figure 2.4    Breakdown of the MBSE ontology*

Figure 2.4 shows that The 'Ontology' is made up of one or more 'Ontology Element' each of which may be classified as one of two types: a 'Cross-cutting Element' or a 'Core Element'.

- A 'Core Element' represents a specific MBSE concept and has its relationships to other Ontology Elements shown on the MBSE Ontology.
- A 'Cross-cutting Element', however, has more complex relationships and may be applied to several basic elements.

Examples of these special types of 'Cross-cutting Element' are shown in Figure 2.5 and are elaborated upon in the subsequent detailed descriptions.

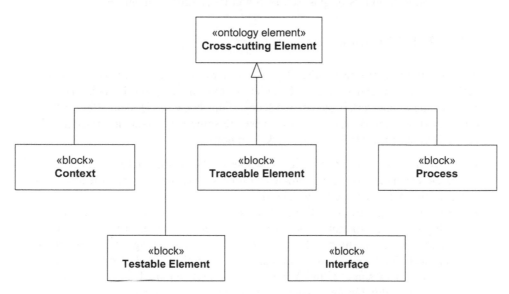

*Figure 2.5    Examples of 'Cross-cutting Element'*

Figure 2.5 shows the concept of Cross-cutting Elements. Some of the concepts that are used in this book have applications across the whole of the MBSE Ontology. Due to the fact that they apply to multiple elements from the MBSE Ontology, they are either not explicitly shown on the MBSE Ontology, or do not have all of their relationships shown, simply because the full MBSE Ontology would become totally unreadable.

- 'Testable Element'. Any Ontology Element in the MBSE Ontology may be tested and, therefore, may be thought of as a Testable Element.
- 'Traceable Element'. Any Ontology Element in the MBSE Ontology may be traced to another Ontology Element in the model and, therefore, may be thought of as a Traceable Element.
- 'Interface Element'. Many Ontology Elements in the MBSE Ontology may have interfaces, such as Process, Stage, View etc. and, therefore, may be thought of as an Interface Element.
- 'Context'. Many Ontology Elements in the MBSE Ontology may have their own Context, and each Pattern in this book has its own Context.

- 'Process'. Many Ontology Elements in the MBSE Ontology have Processes associated with them, such as Need, Life Cycle, Architecture, and, therefore, may be thought of as a Process element.

It should be noted that it is possible for an Ontology Element to be both a Cross-cutting Element and a Core Element, such as Context and Process, whereas others will be one or the other, such as Traceability Element.

The full Ontology is provided in Appendix C. For a full discussion of the MBSE Ontology and its origins, see Holt and Perry (2013, chapter 3).

## 2.3 MBSE Frameworks

A good MBSE approach requires the use of a well-defined Framework that will define the content and structure of the Views that make up the Model. The definition of the Framework is outside the scope of this book, see Holt and Perry (2013) for more discussion on Frameworks. There are some very pragmatic reasons why the definition of the MBSE Framework is important:

- *Coverage*. It is important that the whole of the MBSE Ontology is realised. Each Viewpoint considers a small set of the MBSE Ontology and the totality of the Viewpoints covers the whole MBSE Ontology.
- *Rigour*. By generating all the Views, based on the defined Viewpoints, applying the appropriate Rules for each Viewpoint and ensuring consistency, we produce a true model. This true model provides the rigour for underlying approach. Realising all the Views provide the highest level of rigour, whereas realising only some of the Views provide less rigour. This means that the approach is flexible for projects of different levels of rigour.
- *Approach*. The approach defines how we do things, or to put it another way, the Process that we follow to realise the MBSE Framework by creating Views based on the Viewpoints.
- *Flexibility of scale*. The MBSE Framework defines a number of Viewpoints but, depending on the type of Project being undertaken, not all of these Viewpoints need to be realised as Views. This ability to realise some or all Views makes the MBSE approach very flexible in terms of the size of the Project.
- *Flexibility of realisation*. The Viewpoints defined by the MBSE approach may be realised as Views, each of which may be visualised in any number of different ways. The approach promoted in this book is primarily through using the SysML notation, but any suitable notation may be used to realise the Views. In the same way, any suitable tool may also be used.
- *Integration with other processes*. The MBSE Framework allows integration with any other systems engineering Processes providing that the Information Views for the Processes are known. This allows the MBSE approach to be used with many other methodologies and systems engineering approaches.

- *Automation*. The MBSE Framework provides the basis for automating the MBSE approach using sophisticated systems engineering tools. One of the main benefits of an MBSE approach is that it saves a lot of time and effort as many of the Process Artefacts may be automatically generated.

A Pattern is essentially a special type of Framework therefore all of these points apply equally to both Patterns and Frameworks. The relationship between Patterns and Frameworks will be explored in the next section.

## 2.4   Patterns, Frameworks and the MBSE Ontology

The relationship between Frameworks, Patterns and the MBSE Ontology is shown in Figure 2.6.

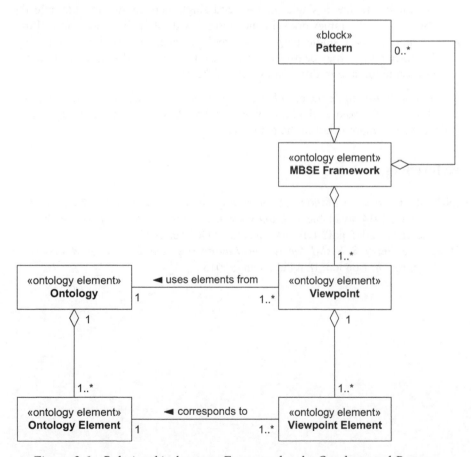

*Figure 2.6   Relationship between Frameworks, the Ontology and Patterns*

Figure 2.6 shows that a 'Pattern' is a special type of 'Framework'. This has several implications:

- As the Pattern is a type of Framework, it means that each Pattern is also made up of one or more Viewpoint, each of which comprises one or more Viewpoint Element.
- The Viewpoints and Viewpoint Elements that comprise the Pattern are also based on the Ontology, providing consistency with the rest of the Model.

Figure 2.6 also shows that a Framework comprises zero or more Pattern. The implication of this is:

- A Framework may be constructed using a number of Patterns.
- A Framework may be constructed using no Patterns.

The main differences between a Framework and a Pattern are:

- A Framework has a single purpose and single application, for example the "seven views" Framework is intended to be used solely for Process modelling.
- A Pattern has a single purpose but multiple applications, for example the Interface Pattern may be used as part of several Frameworks, at different levels of abstraction and in different aspects of the Model.

The approach used to define each Pattern is the same as the approach used to define each Framework and uses the Framework for Architecture Frameworks (FAF) that is described in more detail in the next chapter.

## References

INCOSE *A World in Motion – Systems Engineering Vision 2025,* INCOSE Publishing 2014, available at: http://www.incose.org/docs/default-source/aboutse/se-vision-2025.pdf?sfvrsn=4 [Accessed: October 2015].

Holt, J. and Perry, S. *SysML for Systems Engineering; 2nd Edition: A Model-based Approach*. London: IET Publishing; 2013.

*Chapter 3*

# What is a Pattern?

## 3.1   Introduction

This book is concerned with the application of Patterns in model-based systems engineering. But what is a Pattern and how can they be described in a consistent manner?

This chapter defines what we mean by a Pattern and gives a (very) brief history of the use of Patterns in software and systems engineering. This is followed by a discussion of the key questions that must be considered when defining a Pattern. In order to address these questions the Framework for Architectural Frameworks (FAF) is introduced. The use of the FAF in the definition of Patterns leads naturally to a generic structure for the description of Patterns. This structure is introduced and will be used for all the Pattern definitions in Part II of the book.

## 3.2   What are Modelling Patterns?

In 1977, Christopher Alexander, a professor of architecture at the University of California at Berkeley published an influential book on patterns in architecture (Alexander *et al.*, 1977). In this book Alexander said that a 'pattern describes a problem which occurs over and over again in our environment, and then describes the core of the solution to that problem, in such a way that you can use this solution a million times over, without ever doing it the same way twice.'

Alexander's ideas were adopted by elements of the software engineering object-oriented community in the 1990s before being made more widely known to software engineers through the publication of the so-called *Gang of Four* book in 1995 (Gamma *et al.*, 1995). This book presented a number of design Patterns that could be used to address specific problems in object-oriented software design.

While initially aimed at the object-oriented software community, the idea of Patterns spread to other aspects of software engineering with the publication of books on their use in such areas as analysis (Fowler, 1997) and data modelling (Hay, 1996). Since then the idea of a Pattern – 'an idea that has been useful in one practical context and will probably be useful in others' (Fowler, 1997) – is being adopted into the wider Systems Engineering community. Indeed, many Patterns that were originally created for use in object-oriented software engineering are now being used in Systems Engineering modelling.

An example of this is shown in Figure 3.1. Figure 3.1, taken from (Gamma *et al.*, 1995), shows the composite Pattern that is used 'to represent part-whole hierarchies of objects.' It allows for hierarchies of any depth and width, consisting of 'Composite' objects made up of either other 'Composite' objects or terminal 'Leaf' objects. The example of usage of the Pattern given by Gamma is concerned with the manipulation of graphical objects in drawing editor software.

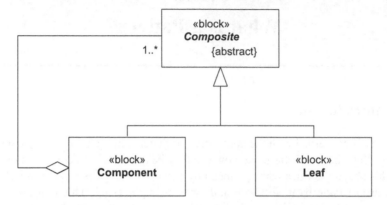

*Figure 3.1    The Composite Pattern*

However, this Pattern is equally applicable outside the software engineering domain. Figure 3.2 shows a typical system hierarchy nomenclature, with a 'System' made up of one or more 'Subsystem' that are made up of one or more 'Assembly' which in turn are made up of one or more 'Subassembly' made up of one or more 'Component'. Finally, each 'Component' is made up of one or more 'Part' which cannot be subdivided further. The diagram has been annotated with *stereotypes* (the words in guillemots, « ») to show which element of the composite Pattern each term is an example of.

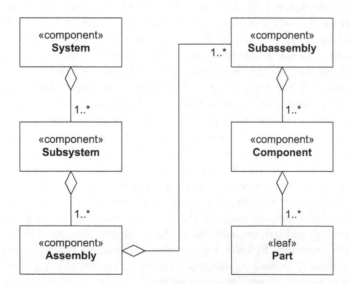

*Figure 3.2    Typical system hierarchy showing stereotypes from the Composite Pattern*

So here we have a Pattern originally described and documented in the context of software engineering design being found to be equally useful in a more general Systems Engineering context. Such re-use has now become wide-spread and the identification, documentation and use of Patterns can significantly aid the Systems Engineering process.

The key concept behind any Pattern, and a concept that must be fully understood in order to get the most benefit from a Pattern, is that a Pattern presents a solution to a type of problem (such as definition of interfaces, how to establish proof of arguments) rather than being a detailed and specific solution to a specific problem. Patterns have to be used with thought and consideration of how (and indeed whether) they can be applied to the problem at hand. They may also require tailoring, but this, again, must be undertaken with thought and consideration in order to ensure that any changes made to the Pattern do not break the Pattern. Major changes may be an indication that a new Pattern has been found.

As has already been alluded to, there are many different kinds of Patterns that can be abstracted and defined, such as analysis Patterns (Fowler, 1997), design Patterns (Gamma *et al.*, 1995) and data modelling Patterns (Hay, 1996). Many organisations are researching and adopting the use of architectural Patterns that describe specific system architectures, both in terms of structure and behaviour. An example of an architectural Pattern would be one that addresses a particular type of control for a system (e.g. centralised vs. distributed). In this book we consider enabling Patterns, specific constructs of modelling elements whose combination and subsequent use enable a number of systems engineering applications. An example of an enabling Pattern would be one used for the definition of interfaces or one used to ensure traceability throughout a model of a system.

## 3.3   Defining Patterns – an introduction to the FAF

One of the cornerstones of MBSE is that of consistency and this is true of Pattern definition just as much as it is of system definition. When defining Patterns for use in MBSE, such definition must be done in such a way that the Pattern is clearly defined using an approach that is consistent across Patterns. As an organisation builds up a library of enabling Patterns, having a consistent approach aids in the documentation, comprehension an adoption of Patterns.

This section introduces the key questions that must be considered when defining a Pattern and outlines a framework-based approach to Pattern definition that ensures that the key questions are addressed in a consistent manner.

### 3.3.1   The key questions

Patterns aren't created out of thin air; they are created to address particular needs, capturing and representing specific concepts in ways that aid in the systems engineering process. In order to approach Pattern definition in a consistent and repeatable way, the following six questions should be considered:

1. **What is the purpose of the Pattern?** It is essential that the purpose and aims of the Pattern are clearly understood so that it can be defined in a concise and consistent fashion.

2. **What concepts must the Pattern support?** All Patterns should be defined around a set of clearly defined concepts that the Pattern is defined to address. For example, a Pattern that addresses issues of proof might contain the concepts of evidence, claim, argument etc.

3. **What different ways of considering the identified concepts are required to fully understand those concepts?** Often it is necessary to consider the concepts covered by a Pattern from a number of different points of view (known in Pattern definition as Viewpoints). A Pattern is defined in terms of the Viewpoints that are needed and so such Viewpoints must be clearly identified and any relationships between them identified. For example, a Pattern that addresses issues of proof might contain Viewpoints that capture claims made, Viewpoints that show supporting arguments etc.

4. **What is the purpose of each Viewpoint?** When Viewpoints have been identified as being needed it is essential that the purpose of each Viewpoint is clearly defined. This is needed for two reasons: firstly, so that the Pattern definer can define the Viewpoint to address the correct purpose and secondly so that anyone using the Pattern understands the purpose of the Viewpoint and can check that they are using the correct Viewpoint (and using the Viewpoint correctly!). Each Viewpoint should have a clear purpose. Good practice also means that each Viewpoint should have, where possible, a single purpose. A common mistake in those with less developed modelling skills is to want to capture all aspects of a problem on a single View. This should be avoided.

5. **What is the definition of each Viewpoint in terms of the identified concepts?** Each Viewpoint is intended to represent a particular aspect of the Pattern and should therefore be defined using only those concepts that have been identified as important for the Pattern. Viewpoints can overlap in the information they present, i.e. concepts can appear on more than one Viewpoint. The Viewpoints must cover all the concepts, i.e. there must be no concept identified as relevant to the Pattern that does not appear on at least one Viewpoint.

6. **What rules constrain the use of the Pattern?** All Patterns are intended to be used to address a specific purpose and so it is important that rules governing the use of the Pattern are defined to aid in its correct use. For example, is there a minimum set of Viewpoints that have to be used for the Pattern to be both useful and use correctly? Are there relationships between Viewpoints that have to be adhered to for the Pattern to be used correctly, for example, are there dependencies between Viewpoints such that information captured on one requires the use of a related Viewpoint? The point of such rules is to ensure consistency. A Pattern is not just a set of unrelated pictures that exist in isolation; when used correctly the result is a set of Views that work together to present the identified concepts in a coherent and consistent fashion.

Having a number of key questions to consider when defining a Pattern is essential. Questions are, however, only as good as their answers. So, in order to ensure that the answers are captured in a consistent fashion, we turn in the following section to a framework-based approach to Pattern definition, the FAF.

### 3.3.2 A (brief) overview of the FAF

The approach to Pattern definition in this book uses the FAF. This section introduces the FAF and shows how the FAF addresses the six key questions. The FAF is defined in detail in Holt & Perry (2013; chapter 11) and is described in more detail, in terms of its usage, in Chapter 19.

The FAF was originally developed to improve the definition of Architectural Frameworks (AFs). However, exactly the same key questions apply to AF definition as to Pattern definition and so the FAF can be used directly for the definition of enabling Patterns as well as AFs. In fact, the FAF is a meta-framework: it is a framework for defining both Patterns and frameworks. The concepts addressed by the FAF are equally applicable to Patterns and frameworks and need the same types of Viewpoints to address them.

The FAF was defined to address the key questions through an MBSE approach, as described in Chapter 2, based around the ideas of Ontology, Viewpoints and Framework:

- Ontology – defines the concepts that are of interest and the relationships between them.
- Viewpoints and Framework – Viewpoints represent the concepts, with each Viewpoint having a defined purpose. Viewpoints can only use concepts from the Ontology and are organised into a Framework.

The FAF consists of an Ontology and six Viewpoints, together with supporting processes. A Viewpoint is a definition of something that is produced when using a Framework or Pattern. The things that are produced are called Views. If it helps, think of a Viewpoint as a recipe and a View as the dish that is created from the recipe. In the following text, abbreviations of the form ABCVp are abbreviations of Viewpoint names. Abbreviations of the form ABCV are abbreviations of View names (i.e. the realisation of a Viewpoint).

Given that the FAF is a meta-framework, you will probably not be surprised to learn that the FAF is defined using the FAF. The six Viewpoints that make up the FAF are shown in Figure 3.3.

The six Viewpoints directly address one of the six key questions:

- 'AF Context Viewpoint' (AFCVp) – What is the purpose of the Pattern? An AFCV defines the context for the Pattern, showing the Pattern concerns in context, establishing why the Pattern is needed.
- 'Ontology Definition Viewpoint' (ODVp) – What concepts must the Pattern support? An ODV defines the ontology for the Pattern. It is derived from the AFCV and defines concepts that can appear on a Viewpoint.

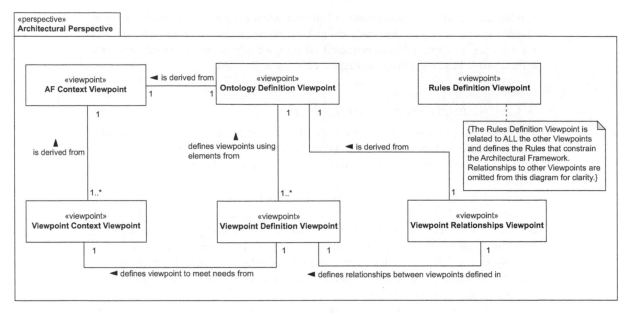

*Figure 3.3    The six FAF Viewpoints*

- 'Viewpoint Relationships Viewpoint' (VRVp) – What different ways of considering the identified concepts are required to fully understand those concepts? A VRV identifies the Viewpoints that are required, shows the relationships between the Viewpoints that make up the Pattern. It is often derived from the ODV.
- 'Viewpoint Context Viewpoint' (VCVp) – What is the purpose of each Viewpoint? Defines the context for a particular Viewpoint. A VCV represents the Viewpoint concerns in context for a particular Viewpoint, establishing why the Viewpoint is needed and what it should be used for. It is derived from the AFCV.
- 'Viewpoint Definition Viewpoint' (VDVp) – What is the definition of each Viewpoint in terms of the identified concepts? A VDV defines a particular Viewpoint, showing the concepts (from the ODV) that appear on the Viewpoint.
- 'Rules Definition Viewpoint' (RDVp) – What rules constrain the use of the Pattern? An RDV defines the various rules that constrain the Pattern.

Thus, when asking each of the key questions one of the FAF Viewpoints should be used to capture the answers to that question. The relationship between the questions and FAF Viewpoints are summarised in Table 3.1.

Because the FAF was initially defined for the definition of AFs, we have a Viewpoint called the AF Context Viewpoint. For Pattern definition, this would better be called the Pattern Context Viewpoint. Feel free to rename the AFCVp to be a PCVp if you wish.

*Table 3.1   How the FAF answers the key questions*

| Key question | FAF Viewpoint used to answer |
|---|---|
| What is the purpose of the Pattern? | Architectural Framework Context Viewpoint (AFCVp) |
| What concepts must the Pattern support? | Ontology Definition Viewpoint (ODVp) |
| What different ways of considering the identified concepts are required to fully understand those concepts? | Viewpoint Relationships Viewpoint (VRVp) |
| What is the purpose of each Viewpoint? | Viewpoint Context Viewpoint (VCVp) – one per Viewpoint being defined |
| What is the definition of each Viewpoint in terms of the identified concepts? | Viewpoint Definition Viewpoint (VDVp) – one per Viewpoint being defined |
| What rules constrain the use of the Pattern? | Rules Definition Viewpoint (RDVp) |

Remember that the FAF defines a number of Viewpoints which are the definitions of something that is produced when using a framework or Pattern. The things that are produced are called Views. So, when using the FAF to define a Pattern, the Pattern definer produces an AF Context View, Ontology Definition View etc. for the Pattern. The Pattern definition will define a number of Viewpoints. For example, the interface Pattern in Chapter 4 defines an Interface Definition Viewpoint. When using the Pattern, an Interface Definition View is produced that conforms to the Interface Definition Viewpoint.

When it comes to realising these Viewpoints (i.e. creating Views based on them during definition of a Pattern) it is suggested that the SysML diagram shown in Table 3.2 are used.

*Table 3.2   Realising the FAF Views in SysML*

| View | SysML diagram used |
|---|---|
| AF Context View (AFCV) | *Use Case Diagram* |
| Ontology Definition View (ODV) | *Block Definition Diagram* |
| Viewpoint Relationships View (VRV) | *Block Definition Diagram** |
| Viewpoint Context View (VCV) | *Use Case Diagram* |
| Viewpoint Definition View (VDV) | *Block Definition Diagram* |
| Rules Definition View (RDV) | *Block Definition Diagram*[†] |

*Use *packages* to show Perspectives.
[†]Although a simple text "diagram" could be used, the use of *blocks* gives us something in the model that can be the source or target of traceability.

More information on the FAF and guidance on using the FAF for Pattern definition are given in Chapter 19 and Appendix D.

Having outlined the FAF and shown how it addresses the key questions related to Pattern definition, we turn, in the following section, to consider how the enabling Patterns in this book are described.

## 3.4    Describing enabling Patterns

Part II defines ten enabling Patterns, each Pattern in its own chapter. All of these enabling Patterns have been defined using the FAF and are described in the same way, as shown in Figure 3.4.

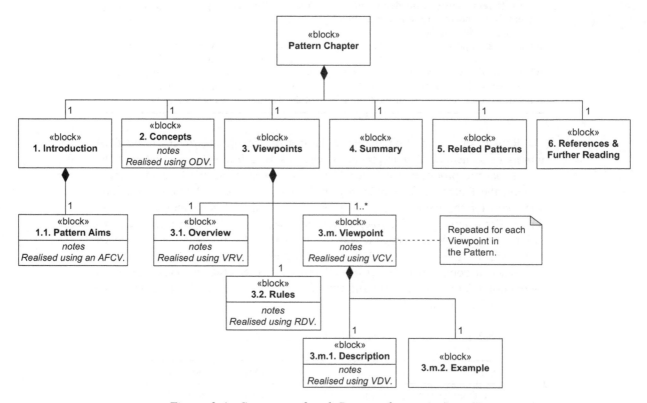

*Figure 3.4    Structure of each Pattern chapter in Part II*

Thus, each enabling Pattern is defined using the following structure:

- **n.1 Introduction** – A brief introduction to the Pattern.
  - ○ **n.1.1 Pattern Aims** – The aims of the Pattern, describing the purpose of the Pattern. The aims are described using an Architectural Framework Context View (AFCV), one of the six FAF Viewpoints.

- **n.2 Concepts** – The concepts that the Pattern addresses, together with the relationships between the concepts. Described using an Ontology Definition View (ODV), one of the six FAF Viewpoints.

- **n.3 Viewpoints** – Identifies and defines the Viewpoints that make up the Pattern. Consists of a number of sub-sections:
  - o **n.3.1 Overview** – Identifies the Viewpoints that make up the Pattern and describes the relationships between the Viewpoints. Described using a Viewpoint Relationships View (VRV), one of the six FAF Viewpoints.
  - o **n.3.2 Rules** – A number of Rules are defined which ensure consistency between the Viewpoints and which define the minimum set of Viewpoints that are needed when using the Pattern. Described using a Rules Definition View (RDV), one of the six FAF Viewpoints.
  - o **n.3.m Viewpoint X** – A description of Viewpoint X of the Pattern. This starts with a statement of the aims that the Viewpoint addresses, described using a Viewpoint Context View, one of the six FAF Viewpoints
    - ■ **n.3.m.1 Description** – Viewpoint X is defined in terms of the Ontology Elements from the Pattern's Ontology (found on its ODV) which can appear on the Viewpoint. This definition of a Viewpoint is described using a Viewpoint Definition View, another of the six FAF Viewpoints.
    - ■ **n.3.m.2 Example** – An example of the Viewpoint in use.
  - o (The n.3.m format is repeated for each Viewpoint in the Pattern.)
- **n.4 Summary** – A summary of the Pattern.
- **n.5 Related Patterns** – Identifies other Patterns in this book that may also be of use when using the Pattern described.
- **n.6 References & Further Reading** – A list of references used in the definition of the Pattern, together with any additional further reading that may be of interest.

When defining your own Patterns it is suggested that you use this structure as a basis for documenting your Patterns. The FAF is used as a framework in which to define the Pattern and the Pattern is then presented in this standard format that makes us of the Views created using the FAF. In this way all Patterns will be defined and documented in a consistent manner, aiding in their understanding and adoption.

# References

Alexander, C.S., Ishikawa, M., Silverstein, M., Jacobson, M., Fiksdahl-Ling, I. and Angel, S. *A Pattern Language.* New York: Oxford University Press; 1977.

Fowler, M. *Analysis Patterns: Reusable Object Models.* Boston, MA: Addison-Wesley; 1997.

Hay, D. *Data Model Patterns: Conventions of Thought.* New York: Dorset House; 1996.

Gamma, E., Helm, R., Johnson, R. and Vlissides, J. *Design Patterns – Elements of Reusable Object Oriented Software.* Boston, MA: Addison-Wesley; 1995.

Holt, J. and Perry, S. *SysML for Systems Engineering; 2nd Edition: A Model-Based Approach.* London: IET Publishing; 2013.

*Part II*

# The fundamental enabling Patterns

*Chapter 4*

# Interface Definition Pattern

## 4.1   Introduction

Interfaces form an integral part of any systems model and define a contract between system elements, whether those elements are physical or are realised in software. They capture the nature of the interactions between those elements, specifying both what can be transferred between the elements and how such transfers take place. Defining interfaces correctly is essential if the system elements are to work properly with each other.

### 4.1.1   Pattern aims

This Pattern is intended to be used as an aid to the definition of interfaces. The main aims of this Pattern are shown in the Architectural Framework Context View (AFCV) in Figure 4.1.

The key aim of the Interface Definition Pattern is to 'Identify Interfaces', the identification of interfaces and their relation to the system elements that use them and the ports that expose them. This Use Case includes further Use Cases:

- 'Define operations' – defining the interfaces in terms of the operations they may provide.
- 'Define flows' – defining the interfaces in terms of the flows of data, material, energy, personnel etc. that take place across an interface.
- 'Identify connections' – identification of the connections between ports that expose interfaces and of the interface connections that take place across those port connections.
- 'Define protocols' – definition of any protocols to which an interface or port must conform.
- 'Identify scenarios' – identification of typical scenarios showing how interfaces are used.

This Pattern can be used for interfaces between physical system elements as well as system elements realised in software. The key Stakeholder Role that benefits from the Interface Definition Pattern is the 'Systems Engineer'.

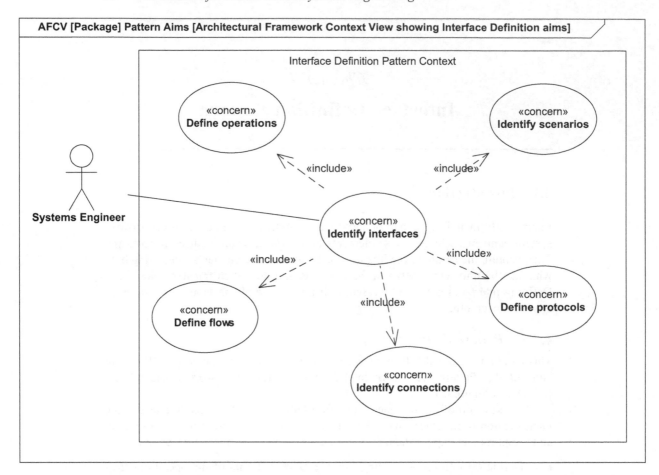

*Figure 4.1    Architectural Framework Context View showing Interface*
*Definition aims*

## 4.2    Concepts

The main concepts covered by the Interface Definition Pattern are shown in the Ontology Definition View (ODV) in Figure 4.2.

Key to this Pattern is the concept of the 'Interface'. An 'Interface' has a 'Direction', which may take the values "in," "out" and "inout." The 'Direction' property of an 'Interface' shows the direction in which the 'Interface' operates from the point of view of the 'Port', owned by a 'System Element', which exposes the 'Interface'.

Two types of 'Interface' exist, the 'Service-Based Interface' and the 'Flow-Based Interface'. Service-Based Interfaces are used to represent those Interfaces that are operation or service based such as are typically found in software-intensive

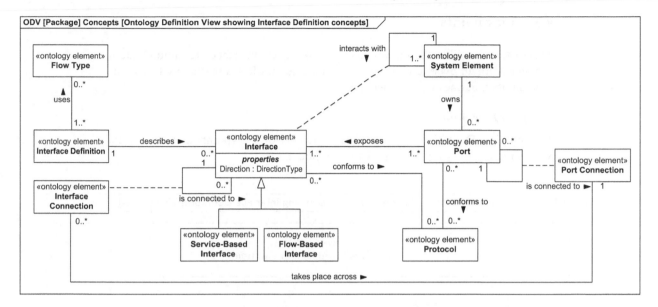

*Figure 4.2   Ontology Definition View showing Interface Definition concepts*

Systems. Flow-Based Interfaces are used to represent those Interfaces that transfer data, material, energy, personnel etc. between System Elements. For example, an Interface between a fuel pump and an engine would be represented by a Flow-Based Interface.

Each 'Interface' is described by an 'Interface Definition'. This defines the operations of a Service-Based Interface and the items transferred by a Flow-Based Interface. These operations and flows use 'Flow Types'. For example, an Interface for a transmitter may have a "transmit" operation that takes a power level parameter. The type of this parameter would be defined using a Flow Type. Similarly, the type of fluid pumped by a pump would be described by a Flow Type.

A 'Port' represents the interaction points between one or more 'System Element' and may represent the concept of a software Port or a physical Port, such as the connector for the fuel line on a car engine fuel pump. Ports are connected to each other via a 'Port Connection'. A fuel rail taking fuel from the fuel pump to fuel injectors in a car engine would be represented by a Port Connection.

Interfaces can be connected together, but only if both ends of the connection are described by the same Interface Definition and have complementary Directions (or if at least one of the ends has Direction "inout"). Such a connection is modelled as an 'Interface Connection' that takes place across a 'Port Connection'. For example, the transfer of fuel from a pump to an engine through a fuel rail would be modelled as an Interface Connection.

Finally, Ports and Interfaces may conform to one or more 'Protocol' that describe and control how the Port and Interface behaves.

## 4.3    Viewpoints

This section describes the Viewpoints that make up the Interface Definition Pattern. It begins with an overview of the Viewpoints, defines Rules that apply to the Pattern and then defines each Viewpoint.

### *4.3.1    Overview*

The Interface Definition Pattern defines a number of Viewpoints as shown in the Viewpoint Relationship View (VRV) in Figure 4.3.

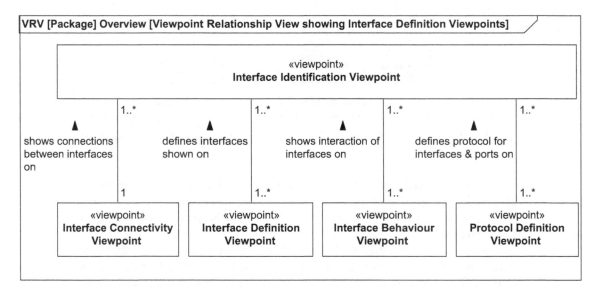

*Figure 4.3    Viewpoint Relationship View showing Interface Definition Viewpoints*

The Interface Definition Pattern defines five Viewpoints for the definition of Interfaces:

- The 'Interface Identification Viewpoint' (IIVp) is used for the identification of Interfaces and Ports and their relation to the System Elements that use them.
- The 'Interface Connectivity Viewpoint' (ICVp) used to show how Interfaces and Ports are connected.
- The 'Interface Definition Viewpoint' (IDVp) is used for the definition of Interfaces in terms of the operations they may provide and the flows of data, material, energy, personnel etc. that take place across an Interface.
- The 'Interface Behaviour Viewpoint' (IBVp) is used for the identification of typical scenarios showing how Interfaces are used.
- The 'Protocol Definition Viewpoint' (PDVp) is used for the definition of any Protocols to which an Interface or Port must conform.

Each of these Viewpoints is described in more detail in the following sections. For each Viewpoint an example is also given.

### 4.3.2   Rules

Six Rules apply to the five IDVps, as shown in the Rules Definition View (RDV) in Figure 4.4.

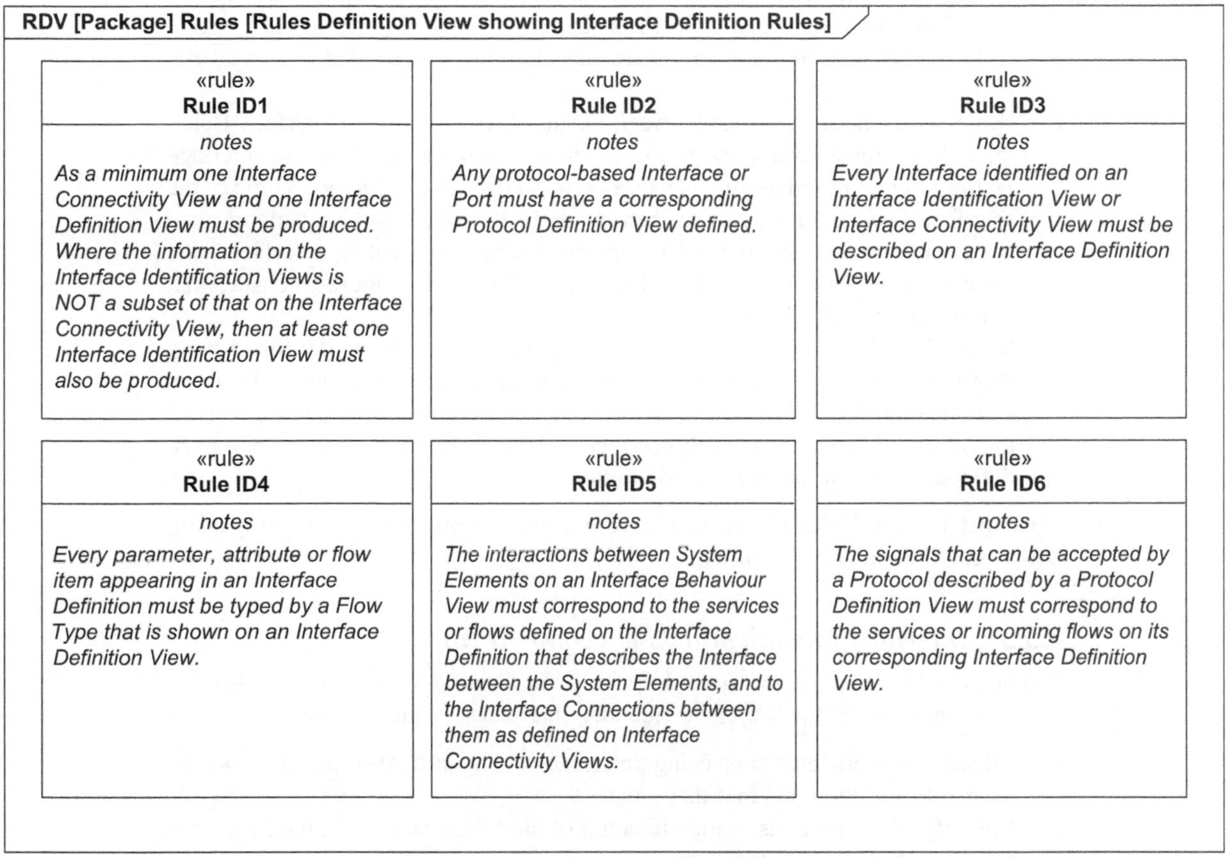

| | | |
|---|---|---|
| **RDV [Package] Rules [Rules Definition View showing Interface Definition Rules]** | | |
| «rule»<br>**Rule ID1**<br><br>*notes*<br><br>*As a minimum one Interface Connectivity View and one Interface Definition View must be produced. Where the information on the Interface Identification Views is NOT a subset of that on the Interface Connectivity View, then at least one Interface Identification View must also be produced.* | «rule»<br>**Rule ID2**<br><br>*notes*<br><br>*Any protocol-based Interface or Port must have a corresponding Protocol Definition View defined.* | «rule»<br>**Rule ID3**<br><br>*notes*<br><br>*Every Interface identified on an Interface Identification View or Interface Connectivity View must be described on an Interface Definition View.* |
| «rule»<br>**Rule ID4**<br><br>*notes*<br><br>*Every parameter, attribute or flow item appearing in an Interface Definition must be typed by a Flow Type that is shown on an Interface Definition View.* | «rule»<br>**Rule ID5**<br><br>*notes*<br><br>*The interactions between System Elements on an Interface Behaviour View must correspond to the services or flows defined on the Interface Definition that describes the Interface between the System Elements, and to the Interface Connections between them as defined on Interface Connectivity Views.* | «rule»<br>**Rule ID6**<br><br>*notes*<br><br>*The signals that can be accepted by a Protocol described by a Protocol Definition View must correspond to the services or incoming flows on its corresponding Interface Definition View.* |

*Figure 4.4   Rules Definition View showing Interface Definition Rules*

The six Rules are:

- 'Rule ID1: As a minimum one Interface Connectivity View (ICV) and one Interface Definition View (IDV) must be produced. Where the information on the Interface Identification Views (IIVs) is NOT a subset of that on the ICV, then at least one IIV must also be produced.' This Rule establishes the minimum set of Views that must be considered when using the Pattern. If this Rule is not followed then an incomplete definition of Interfaces will result.

- 'Rule ID2: Any protocol-based Interface or Port must have a corresponding Protocol Definition View defined.' Again, this Rule is ensuring completeness of the definition of Interfaces.
- 'Rule ID3: Every Interface identified on an IIV or ICV must be described on an IDV.' This Rule ensures that Interfaces are defined in terms of their Flow Types and behaviours so that undefined Interfaces cannot be used in the Interface definition.
- 'Rule ID4: Every parameter, attribute or flow item appearing in an Interface Definition must be typed by a Flow Type that is shown on an IDV.' Like Rule 3, this Rule is there to ensure that nothing is used that has not been defined.
- 'Rule ID5: The interactions between System Elements on an Interface Behaviour View must correspond to the services or flows defined on the Interface Definition that describes the Interface between the System Elements, and to the Interface Connections between them as defined on ICVs.' This Rule ensures that, when considering the behaviour and interaction through Interfaces, the interactions shown have been defined; again this Rule is there to ensure that nothing is used that has not been defined.
- 'Rule ID6: The signals that can be accepted by a Protocol described by a Protocol Definition View must correspond to the services or incoming flows on its corresponding IDV.' Again, this Rule is there to ensure that nothing is used that has not been defined. A Protocol cannot use signals unless they have been defined as part of the Interface definition.

Note that the six Rules shown in Figure 4.4 are the minimum that are needed. Others could be added if required.

### 4.3.3   Interface Identification Viewpoint (IIVp)

The aims of the IIVp are shown in the Viewpoint Context View in Figure 4.5.
    The main aim of the IIVp is to 'Identify Interfaces', which includes:

- 'Identify system elements using interfaces' – identification of the System Elements and the Ports that they own.
- 'Identify exposing ports' – identification of the Ports that expose the Interfaces and the Interfaces that they expose.

The key Stakeholder Role that benefits from the 'Identify interfaces' Use Case is the 'Systems Engineer'.

#### 4.3.3.1   Description

The Viewpoint Definition View (VDV) in Figure 4.6 shows the Ontology Elements that appear on an IIVp.
    The IIVp shows System Elements and the Ports that they own. The Interfaces exposed by Ports are also shown, along with the names of the Interface Definitions that describe them.

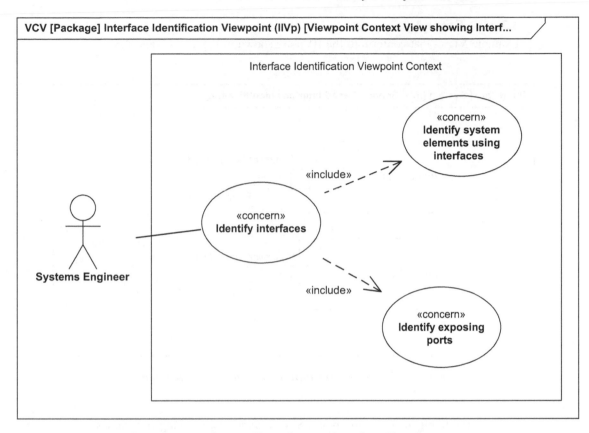

*Figure 4.5   Viewpoint Context View showing Interface Identification Viewpoint aims*

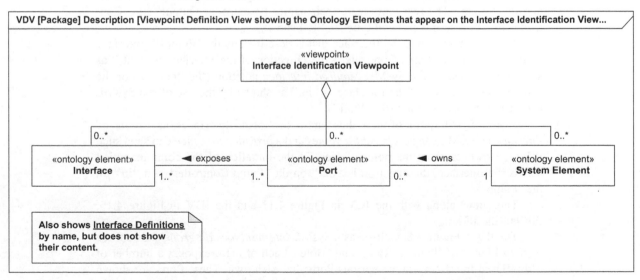

*Figure 4.6   Viewpoint Definition View showing the Ontology Elements that appear on the Interface Identification Viewpoint (IIVp)*

#### 4.3.3.2   Example

Example Views that conform to the IIVp are shown in Figures 4.7 and 4.8.

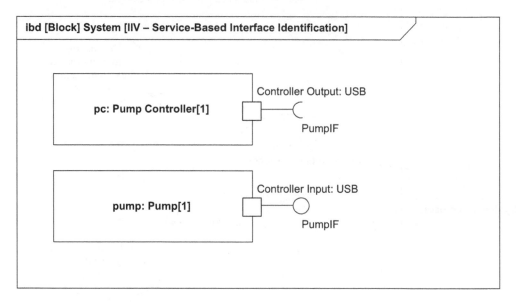

*Figure 4.7   IIV – Service-Based Interface Identification*

The IIV in Figure 4.7, realised as a SysML *internal block diagram*, shows two System Elements, namely a 'Pump Controller' and a 'Pump'. Each of these has a Port shown by the small squares on the right-hand edges of the 'Pump Controller' and the 'Pump'. The two Ports have their names ('Controller Output' and 'Controller Input', respectively) and type ('USB') shown.

The Ports both expose an Interface that is described by the 'PumpIF' Interface Definition. For the 'Pump Controller', the Direction of the Interface is "out," as shown by the use of the SysML *required interface* notation (the "cup"). For the 'Pump', the Direction of the Interface is "in," as shown by the use of the SysML *provided interface* notation (the "ball").

From a SysML point of view these are both "inout" Interfaces, as a *required interface* in SysML can accept *return values* and a *provided interface* can send such values. However, from the point of view of the initiation of the communication across the Interface, the direction is "out" on the 'Pump Controller' and "in" on the 'Pump'.

This view, along with the ICV in Figure 4.11 and the IDV in Figure 4.15, fulfils Rule ID1.

The IIV in Figure 4.8, realised as a SysML *internal block diagram*, shows three System Elements: 'Pump', 'Tank' and 'Hole'. Each of these exposes a number of Flow-Based Interfaces via the various Ports that each own. These Ports are shown

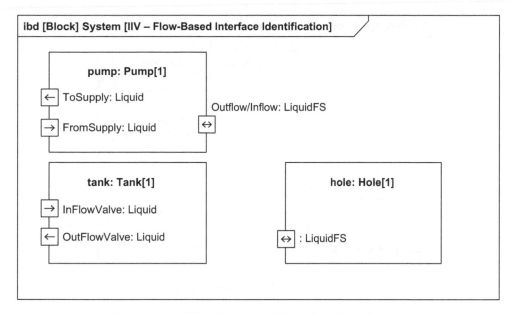

*Figure 4.8   IIV – Flow-Based Interface Identification*

using SysML *flow ports*, the small squares containing arrow heads, which indicate the directionality of the Interfaces.

The 'Pump' has an *out* Flow-Based Interface named 'ToSupply' that can transfer 'Liquid', an "in" Flow-Based Interface named 'FromSupply' that can receive 'Liquid' and an "inout" Flow-Based Interface named 'Outflow/Inflow' that is of type 'LiquidFS'.

The 'Tank' has an "in" Flow-Based Interface named 'InFlowValve' that can receive 'Liquid' and an "out" Flow-Based Interface named 'OutFlowValve' that can transfer 'Liquid'.

Finally, the Interface to the 'Hole' is modelled as an un-named "inout" Flow-Based Interface that is of type 'LiquidFS'.

This View, along with the ICV in Figure 4.12 and the IDV in Figure 4.15, fulfils Rule ID1.

### 4.3.4   Interface Connectivity Viewpoint (ICVp)

The aims of the ICVp are shown in the Viewpoint Context View in Figure 4.9.

The main aim of the ICVp is to 'Identify Interfaces' with an emphasis on 'Identify Connections', showing the connections between Ports and the Interface Connections that take place across the Port Connections. The key Stakeholder Role that benefits from the 'Identify interfaces' Use Case is the 'Systems Engineer'.

#### 4.3.4.1   Description

The VDV in Figure 4.10 shows the Ontology Elements that appear on an ICVp.

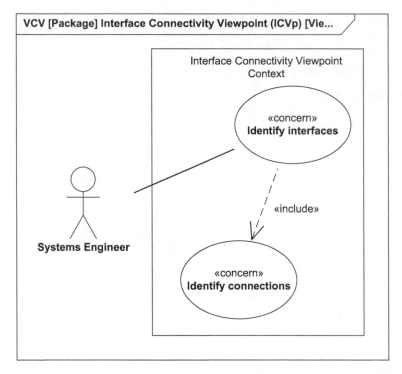

*Figure 4.9    Viewpoint Context View showing Interface Connectivity Viewpoint aims*

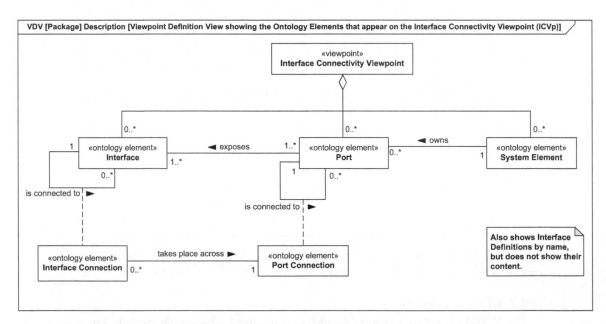

*Figure 4.10    Viewpoint Definition View showing the Ontology Elements that
appear on the Interface Connectivity Viewpoint (ICVp)*

The ICVp shows System Elements and the Ports that they own. The Port Connections between Ports are shown together with the Interfaces exposed by Ports and the Interface Connections between these Interfaces.

The names of the Interface Definitions that describe each Interface are shown, but the content of the Interface Definitions is not shown.

The ICVp can be thought of as containing the IIVp. However, not all the Ports and Interfaces shown on an IIV (an instance of an IIVp) need be connected, or, alternatively, they can be connected in different ways depending on different configurations of System Elements. In such cases then both diagrams are needed, otherwise the IIVp can be omitted.

#### 4.3.4.2 Example

Example Views that conform to the ICVp are shown in Figures 4.11 and 4.12.

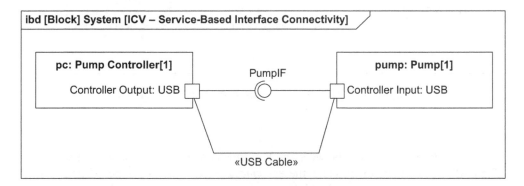

*Figure 4.11   ICV – Service-Based Interface Connectivity*

The ICV in Figure 4.11, realised as a SysML *internal block diagram*, shows two System Elements, namely a 'Pump Controller' and a 'Pump'. Each of these has a Port shown by the small squares on the right and left edges of the 'Pump Controller' and the 'Pump', respectively. The Port Connection between these two Ports is shown, and use has been made of SysML *stereotypes* to annotate the *connector* to show that physically this is a USB cable as shown by the «USB Cable» *stereotype*.

Both Ports expose an Interface that is described by the 'PumpIF' Interface Definition. The Interface Connection between them is shown by the connection of the SysML "cup" and "ball" notation used to indicate the Interfaces.

This View, along with Figures 4.7 and 4.15, fulfils Rule ID1.

The ICV in Figure 4.12, realised as a SysML *internal block diagram*, identifies three Flow-Based Interfaces between three System Elements. There are two between the 'Tank' and the 'Pump' and one between the 'Pump' and the 'Hole'.

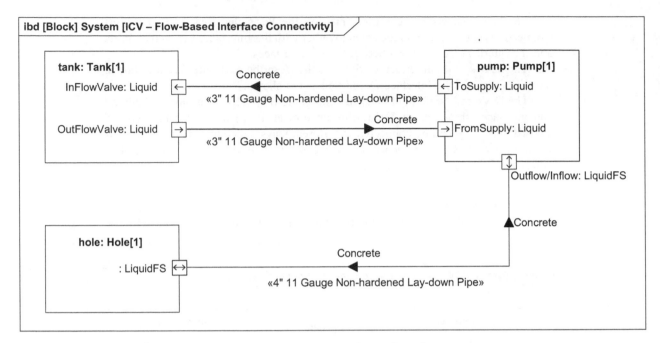

*Figure 4.12    ICV – Flow-Based Interface Connectivity*

The Port Connections between the Ports are shown, and use has been made of SysML *stereotypes* to annotate these *connectors* to show the type of physical connection. There are two '3″ 11 Gauge Non-hardened Lay-down Pipe' between the 'Tank' and the 'Pump' and one '4″ 11 Gauge Non-hardened Lay-down Pipe' between the 'Pump' and the 'Hole'.

The SysML *flow ports* used on the diagram show the directionality of the Interfaces. The Interfaces between the 'Pump' and the 'Hole' have a Direction of "inout". The 'Pump' can pump into and out of the 'Hole' through a single pipe. The two Interfaces between the 'Tank' and the 'Pump' are each uni-directional. From the point of view of the 'Tank' it has an Interface with a Direction of "out" to supply 'Concrete' to the 'Pump' via its 'OutFlowValve', and an Interface with a Direction of "in" to receive 'Concrete' from the 'Pump' via its 'InFlowValve'. The SysML *item flows* carrying 'Concrete' across the *connectors* define the Interface Connections.

This View, along with Figures 4.8 and 4.15, fulfils Rule ID1.

### 4.3.5    *Interface Definition Viewpoint (IDVp)*

The aims of the IDVp are shown in the Viewpoint Context View in Figure 4.13.

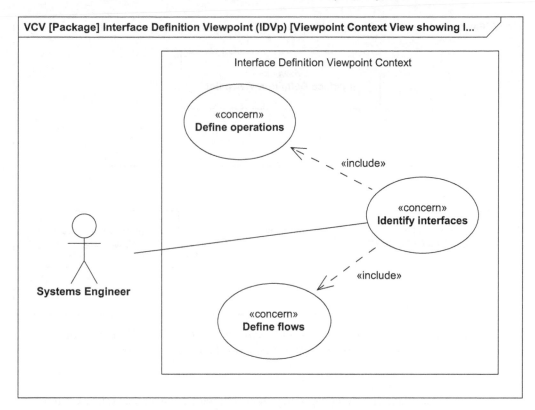

*Figure 4.13    Viewpoint Context View showing Interface Definition Viewpoint aims*

The aim of the IDVp is to 'Identify interfaces', with an emphasis on:

- 'Define operations' – definition of Interfaces in terms of the operations they provide.
- 'Define flows' – definition of Interfaces in terms of the flows of data, material, energy, personnel etc. that take place across an Interface.

The key Stakeholder Role that benefits from the 'Identify interfaces' Use Case is the 'Systems Engineer'.

### 4.3.5.1    Description
The VDV in Figure 4.14 shows the Ontology Elements that appear on an IDVp.

The IDVp contains a number of Interface Definitions, together with the Flow Types of the items passed across the Interfaces that are described by the Interface Definitions.

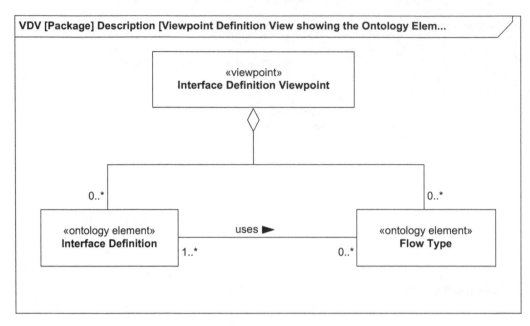

*Figure 4.14    Viewpoint Definition View showing the Ontology Elements that appear on the Interface Definition Viewpoint (IDVp)*

Note that this Viewpoint does not show Interfaces, but concentrates on the descriptions of Interfaces through the Interface Definitions and Flow Types that describe them.

### 4.3.5.2   Example

An example View that conforms to the IDVp is shown in Figure 4.15.

This IDV, here realised as a SysML *block definition diagram*, shows three Interface Definitions.

The first Interface Definition is 'PumpIF' modelled using a SysML *interface block*. 'PumpIF' is an example of an Interface Definition for a Service-Based Interface. It defines a number of services that the Interface provides, realised here using SysML *operations*, an example of which is the 'start' service. It also defines a single *property*, 'CurrentDirection', which is used to store information about the state of the Interface when it is in use.

'DirectionType', 'Boolean' and 'PowerLevel' are all examples of Flow Types that are used by the 'PumpIF' as the types of *parameters* of the three services 'start', 'stop' and 'reverse' and of the 'CurrentDirection' *property*. The "uses" relationship is made explicit through the use of the stereotyped SysML *dependency*.

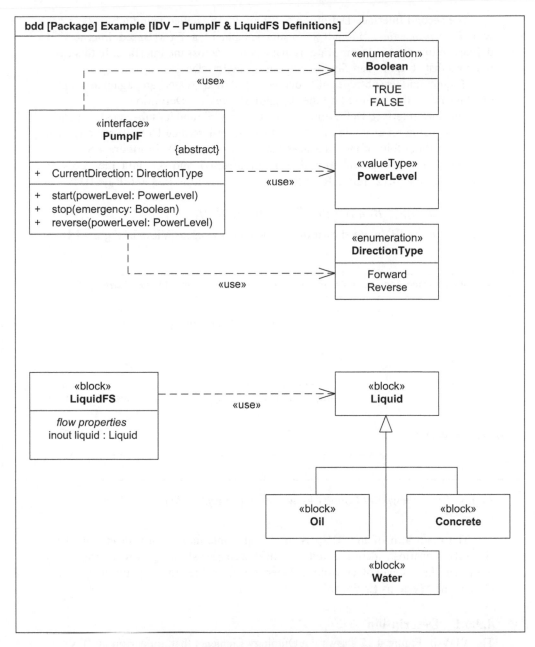

*Figure 4.15　IDV – PumpIF and LiquidFS Definitions*

The second Interface Definition is 'LiquidFS' modelled using a SysML *block* with *flow properties*. 'LiquidFS' is an example of a Flow-Based Interface. It defines an Interface in terms of items that can flow across the Interface. In this case, it shows that 'Liquid' can flow in and out of the Interface.

'Liquid' and its sub-types of 'Concrete', 'Oil' and 'Water' are, again, examples of Flow Types that are used by the 'LiquidFS' Interface Definition.

The third Interface Definition is given by 'Liquid' (and its sub-types). 'Liquid', as well as being an example of a Flow Type, is an Interface Definition in its own right, describing four of the Interfaces that appear on the IIV in Figure 4.8.

This View, along with the IIVs in Figures 4.7 and 4.8 and the ICVs in Figures 4.11 and 4.12, fulfils Rule ID1. It also fulfils Rule ID3 and Rule ID4.

### 4.3.6    Interface Behaviour Viewpoint (IBVp)

The aims of the IBVp are shown in the Viewpoint Context View in Figure 4.16.

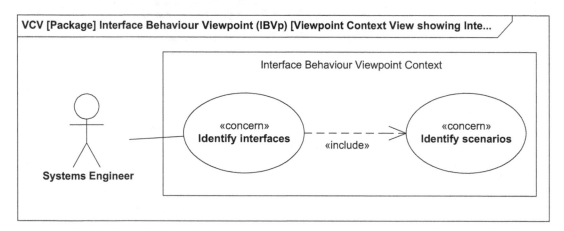

*Figure 4.16    Viewpoint Context View showing Interface Behaviour Viewpoint aims*

The main aim of the IBVp is to 'Identify interfaces' with an emphasis on 'Identify scenarios', identification of typical scenarios showing how interfaces are used. The key Stakeholder Role that benefits from the 'Identify interfaces' Use Case is the 'Systems Engineer'.

#### 4.3.6.1    Description

The VDV in Figure 4.17 shows the Ontology Elements that appear on an IBVp.

The IBVp shows a number of System Elements interacting with each other via the services and items transferred across the Interfaces between the System Elements. Since these interactions are governed by the Interface Definitions and associated Flow Types that describe each Interface, the IBVp indirectly shows elements of the Interface Definitions and Flow Types.

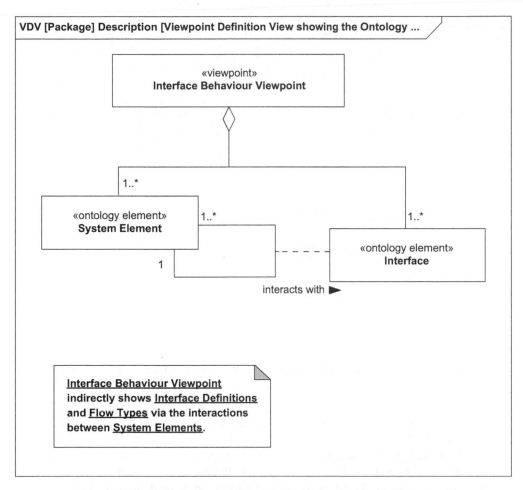

*Figure 4.17   Viewpoint Definition View showing the Ontology Elements that appear on the Interface Behaviour Viewpoint (IBVp)*

### 4.3.6.2   Example

Example Views that conform to the IBVp are shown in Figures 4.18–4.20.

The Interface Behaviour View in Figure 4.18, here realised as a SysML *sequence diagram*, shows the interactions between two System Elements, the 'Pump Controller' and the 'Pump'.

As shown in Figure 4.11, all interactions between the 'Pump Controller' and the 'Pump' must conform to 'PumpIF'. That is, they must conform to the Interface Definition that is described on the IDV in Figure 4.15. The diagram shows *messages* corresponding to the 'start', 'stop' and 'reverse' services shown on the IDV being sent from the 'Pump Controller' to the 'PumpIF'. The 'Controller Input' *port* of type 'USB' that exposes the 'PumpIF' is also shown.

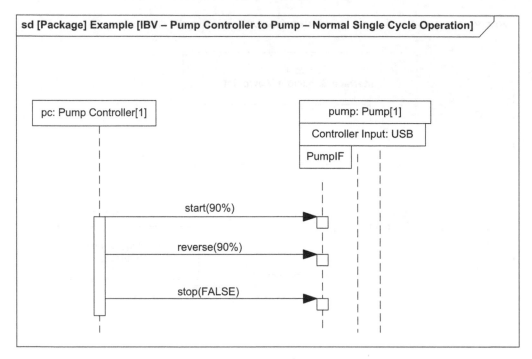

*Figure 4.18   IBV – Pump Controller to Pump – Normal Single Cycle Operation*

This View, along with the IDV in Figure 4.15 and the ICV in Figure 4.11, fulfils Rule ID5.

This example illustrates a simple Interface without any governing Protocol. An example of an Interface Behaviour View for the 'Pump Controller' and 'Pump' where the 'PumpIF' does conform to a governing Protocol can be seen in Figure 4.19.

The Interface Behaviour View in Figure 4.19, here realised as a SysML *sequence diagram*, shows two System Elements, the 'Pump Controller' and the 'Pump', connected together by the 'PumpIF', an example of a Service-Based Interface.

As shown in Figure 4.11, all interactions between the 'Pump Controller' and the 'Pump' must conform to 'PumpIF'. That is, they must conform to the Interface Definition that is described on the IDV in Figure 4.15. This can be seen in the diagram where SysML *messages* corresponding to the 'start', 'stop' and 'reverse' services shown on the IDV can be seen being sent from the 'Pump Controller' to the 'PumpIF'.

In this example the 'PumpIF' has an internal Protocol (see Figure 4.23) that translates the 'start', 'stop' and 'reverse' *messages* that it receives into a series of 'prime', 'pump', 'stopPump', 'pumpReverse' and 'flush' *signals* that it forwards to the 'Pump'.

Although the 'PumpIF' is shown as a separate SysML *lifeline* in the diagram, this has been done to emphasise the internal behaviour of the Interface and to make explicit the translations that are performed by its governing Protocol. The 'PumpIF' should not

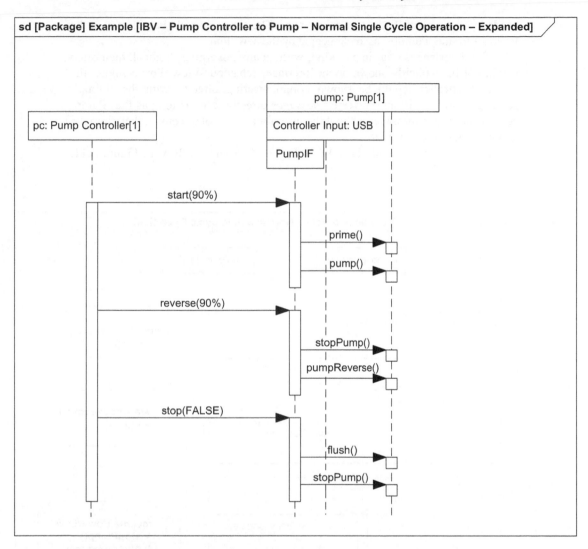

*Figure 4.19   IBV – Pump Controller to Pump – Normal Single Cycle*
*Operation – Expanded*

be thought of as being separate from the 'Pump'; it simply defines a set of services provided by the 'Pump'. When implementing this System, this aspect of the behaviour of 'PumpIF' could be implemented in software running on the 'Pump'. The 'PumpIF' is acting as a "wrapper" to the 'Pump', providing a simple set of three services that can remain constant if the internal operation of the 'Pump' changes or that can be used on pumps with different behaviour. For example, if a self-priming pump that could go directly from pumping to pumping in reverse without stopping was to replace the existing 'Pump', then provided it exposed the same 'PumpIF' no changes would be needed to the type of 'Pump Controller' used. The changes would be reflected in the Protocol for the 'PumpIF' as implemented by the new 'Pump'.

If an Interface is of a simpler kind, without any governing Protocol, then often it will not be explicitly shown on an Interface Behaviour View. For example, the 'Pump Controller' could be shown communicating directly with the 'Pump', sending 'start', 'stop' and 'reverse' *messages* directly to it, as long as the 'Pump' could handle such messages without the need for a Protocol to convert them into the 'prime', 'pump' etc. *signals*.

This View, along with the IDV in Figure 4.15 and the ICV in Figure 4.11, fulfils Rule ID5.

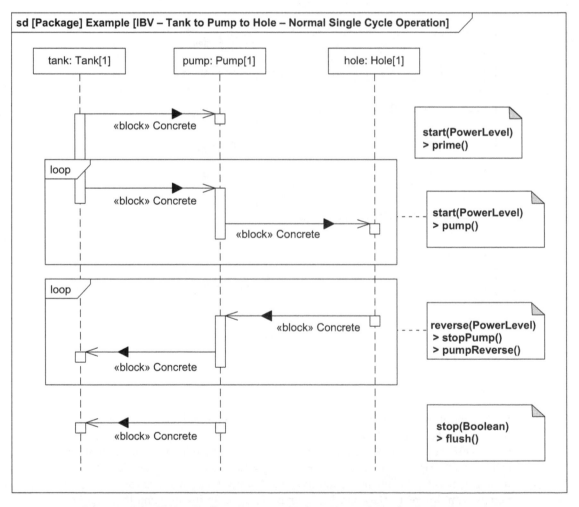

*Figure 4.20   IBV – Tank to Pump to Hole – Normal Single Cycle Operation*

The Interface Behaviour View in Figure 4.20, modelled here using a SysML *sequence diagram*, shows an example of System Elements interacting via Flow-Based Interfaces.

The diagram shows three System Elements: 'Tank', 'Pump' and 'Hole'. As shown in Figure 4.12, they interact according to the 'LiquidFS' and 'Liquid' Interface Definitions that are defined on the IDV in Figure 4.15. The scenario shown in the View corresponds to that shown in the corresponding Service-Based Interface IBV seen in Figure 4.19, but from the point of view of the items flowing between the various System Elements rather than the services invoked by one on another.

It should also be noted that, in this scenario, it is 'Concrete' that is flowing between the various System Elements. Because of the way the Interfaces are defined any of the defined sub-types of 'Liquid', such as 'Concrete', 'Oil' or 'Water' could have been used.

Also, although the behaviour of the two types of Interface has been shown on separate Interface Behaviour Views, there is nothing to prevent these two diagrams being combined in to a single IBV showing the behaviour of both types on a single diagram.

This View, along with the IDV in Figure 4.15 and the ICV in Figure 4.12, fulfils Rule ID5.

### 4.3.7  Protocol Definition Viewpoint (PDVp)

The aims of the PDVp are shown in the Viewpoint Context View in Figure 4.21.

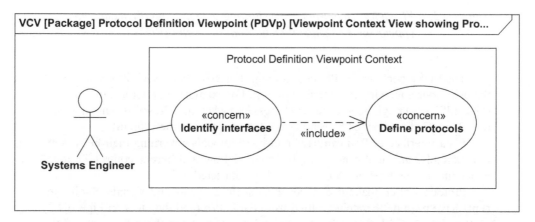

*Figure 4.21   Viewpoint Context View showing Protocol Definition Viewpoint aims*

The main aim of the PDVp is to 'Identify interfaces' with an emphasis on 'Define protocols', defining any protocols to which an interface or port must conform. The key Stakeholder Role that benefits from the 'Identify interfaces' Use Case is the 'Systems Engineer'.

#### 4.3.7.1   Description

The VDV in Figure 4.22 shows the Ontology Elements that appear on a PDVp.

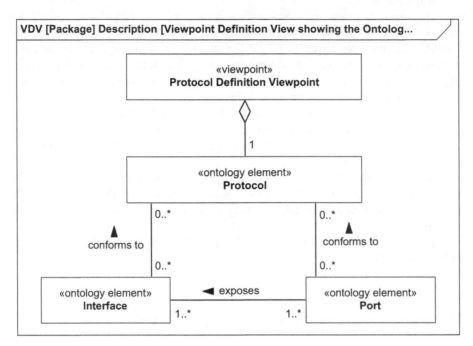

*Figure 4.22    Viewpoint Definition View showing the Ontology Elements that appear on the Protocol Definition Viewpoint (PDVp)*

The PDVp contains the Protocol for an Interface or Port. Such Protocols define the behaviour governing the Interface or Port. For example, an interface to a 'Pump' ('PumpIF', say) might ignore 'reverse' messages when the 'Pump' is pumping until the 'Pump' is first stopped by sending a 'stop' message to the 'PumpIF'.

Each Interface or Port can have multiple Protocols governing their behaviour. For example, an intelligent 'PumpIF' could follow a different control Protocol depending on the type of 'Pump Controller' connected to it.

Protocols often make use of concepts such as events and signals. Such concepts have been deliberately omitted from the PDVp (and the ODV in Figure 4.2) as they are dependent on the representation adopted for the realisation of the Viewpoint.

#### 4.3.7.2   Example

An example View that conforms to the PDVp is shown in Figure 4.23.

The Protocol Definition View, here realised as a SysML *state machine diagram*, describes the Protocol to which the 'PumpIF' must conform.

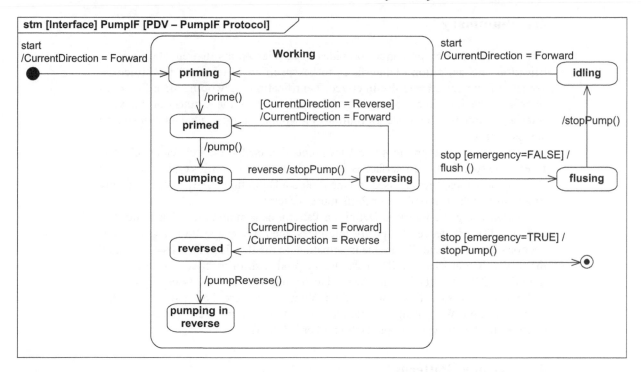

*Figure 4.23    PDV – PumpIF Protocol*

The 'PumpIF' provides three services: 'start', 'stop' and 'reverse'. It must convert invocations of these services into the relevant *signals* to be issued to the 'Pump' to which it provides an Interface.

In this example, an invocation of the 'start' service must be converted into a 'prime' and then a 'pump' *signal* to the 'Pump'. Similarly, 'reverse' is converted into 'stopPump' and 'pumpReverse' *signals* and 'stop' into either a 'flush' and then a 'stopPump' *signal* or just a 'stopPump' *signal* depending on whether an emergency stop is being requested.

In order to be able to correctly handle a 'reverse' service request, the 'PumpIF' must maintain information on which direction the 'Pump' is running. This is held in the 'CurrentDirection' *property* which is given the value 'Forward' or 'Reverse' as appropriate.

An observation that can be made about this diagram is that it takes no account of the 'powerLevel' *parameter* passed in with the 'start' and 'reverse' service calls (and shown on the Interface Behaviour View in Figure 4.19. Perhaps this implementation of the 'PumpIF' Protocol is intended to be used with a 'Pump' that does not take a 'powerLevel'. If it does take a 'powerLevel' then the handling of 'powerLevel' should be added to this View and this would also require changes to the Interface Behaviour View in Figure 4.19.

This View fulfils Rule ID2 and Rule ID6.

## 4.4    Summary

The Interface Definition Pattern provides three Viewpoints that enable the identification and definition of Interfaces to be specified in terms of the structural aspects of the Interfaces: the Interface Identification Viewpoint identifies each Interface, the Interface Connectivity Viewpoint shows the connection between Interfaces and the Interface Definition Viewpoint defines what is transferred across each Interface.

The Pattern also provides two Viewpoints that enable the behaviour of Interfaces to be specified: the Interface Behaviour Viewpoint identifies typical scenarios showing how Interfaces are used and the Protocol Definition Viewpoint defines any Protocols to which Interfaces or Ports must conform.

When using the Interface Definition Pattern, as a minimum at least one ICV and one IDV are needed to specify Interfaces, their associated Ports and the connections between them. Where the information on the IIVs is not a subset of that on the ICVs, then at least one IIV must also be produced. In practice, however, multiple IIVs, ICVs and IDVs would be produced along with Interface Behaviour View and, where necessary, Protocol Definition Views. Note here the use of View rather than Viewpoint. When using the Interface Definition Pattern, Views are created that conform to the Viewpoints (see Holt and Perry (2013)).

## 4.5    Related Patterns

If using the Interface Definition Pattern, the following Patterns may also be of use:

- Description
- Traceability.

## Reference

Holt, J. and Perry, S. *SysML for Systems Engineering; 2nd Edition: A Model-Based Approach*. London: IET Publishing; 2013.

## Further readings

Object Management Group. *SysML website* [online]. Available at: http://www.omgsysml.org [Accessed February 2015].

Holt, J. *UML for Systems Engineering – Watching the Wheels*. 2nd Edition. London: IET Publishing; 2004 (reprinted IET Publishing; 2007).

Rumbaugh, J., Jacobson, I. and Booch, G. *The Unified Modeling Language Reference Manual*. 2nd Edition. Boston, MA: Addison-Wesley; 2005.

Stevens, R., Brook, P., Jackson, K. and Arnold, S. *Systems Engineering – Coping with Complexity*. London: Prentice Hall Europe; 1998.

## Chapter 5
# Traceability Pattern

## 5.1 Introduction

There are many interpretations of traceability, including:

- Measurement
- Logistics
- Materials
- Supply Chain
- Development.

However, two themes run throughout all interpretations: an unbroken chain and access to information. The unbroken chain focuses on the need to be able to follow an artefact back to its source, whether that source is a physical place such as the sea, a supplier organisation, a requirement, meeting minutes or a standard such as ISO 15288. Access to information defines the need to know about the artefact being traced. In some cases this may be the history of the artefact. For example, in the case of food, this could be the organisations the food has passed through, when and who did the quality check etc. For an engineering project, it may be why an artefact exists or has been chosen as the solution to a requirement.

The existence of traceability supports areas of analysis including: impact analysis, change management and coverage analysis. All of these types of analysis would be difficult if not impossible to carry out without traceability in place.

### 5.1.1 Pattern aims

This Pattern is intended to be used as an aid to the use of traceability within a Model-Based Systems Engineering (MBSE) application. The main aims of this Pattern are shown in the Architectural Framework Context View (AFCV) in Figure 5.1.

The key aim of the Traceability Pattern is to allow a 'Systems Engineer' to 'Establish traceability' in a model of a System. This key aim is constrained by the need to:

- 'Support traceability throughout model' – Support the capture of the traceability relationships between traceable elements in a systems engineering model, throughout the model and between any desired model elements.
- 'Support impact analysis' – Support 'Systems Engineers' and 'Systems Engineering Managers' in the identification of model elements that may be impacted by change.

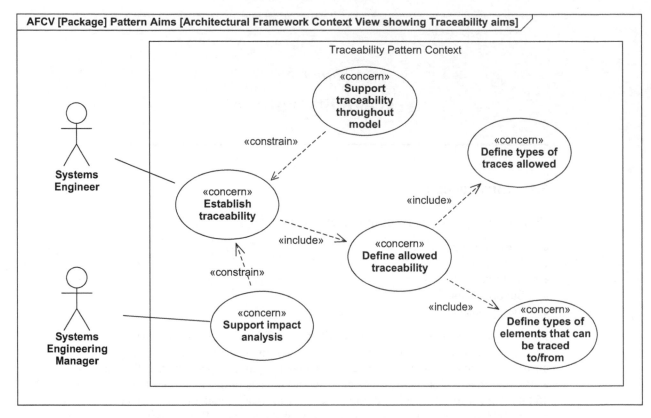

*Figure 5.1   Architectural Framework Context View showing Traceability aims*

The main aim of 'Establish traceability' includes the aim of 'Define allowed traceability'. This in turn includes the uses cases:

- 'Define types of trace allowed' – Support the definition of the types of traces that can be used.
- 'Define types of element that can be traced to/from' – Support the definition of the types of elements that can be involved in trace relationships and the relationships that can be used between such traceable elements.

The Use Case 'Establish traceability' provides benefit to the 'Systems Engineer' Stakeholder Role. The 'Systems Engineering Manager' Stakeholder Role gains benefit from the 'Support impact analysis' Use Case.

## 5.2   Concepts

The main concepts covered by the Traceability Pattern are shown in the Ontology Definition View (ODV) in Figure 5.2.

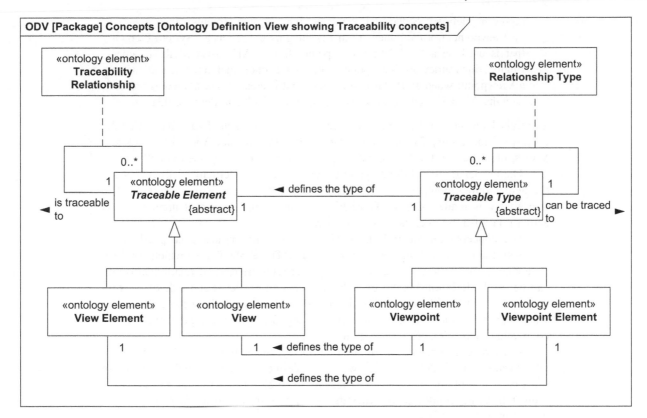

*Figure 5.2    Ontology Definition View showing Traceability concepts*

Key to getting traceability right is establishing the understanding of the way in which traceability will be used. This means understanding the types of information to be traced and the relationships that can be used to realise the traceability.

The right-hand side of this diagram shows the types of things which can be traced – a 'Traceable Type'. These may be:

- 'Viewpoint', representing types of 'View' occurring in a model. 'Viewpoint' may represent underlying diagram types from the modelling language being used, such as a *block definition diagram* if using SysML. They may also represent defined Viewpoints from a framework that is being used, such as a 'Validation Viewpoint' from a Model-Based Requirements Engineering framework. Viewpoints are conceptual in nature; they are a definition to which an instance (a View) must conform. See Holt and Perry (2013).
- 'Viewpoint Element', representing types of 'View Element' occurring in a model. Viewpoint Elements may represent underlying modelling element types from the modelling language being used, such as a *block* if using SysML. They may also

represent defined conceptual elements (Ontology Elements) that are used on Viewpoints from a framework that is being used, such as a "System Element" that is used on a "Validation Viewpoint" from a Model-Based Requirements Engineering framework. Viewpoint Elements are conceptual in nature and make up a Viewpoint; when an instance of a Viewpoint is created (i.e. a View is created), then that View is made up of View Elements. See Holt and Perry (2013).

The right-hand side of the diagram also includes 'Relationship Type' that define the types of 'Traceability Relationship' that can be used to trace View Elements and Views, identified by the Viewpoints and Viewpoint Elements, one to another.

The left-hand side of the diagram shows 'Traceable Element'. This represents the actual things being traced between, i.e. the 'View' or 'View Element'. The traceability is made using a 'Traceability Relationship', a representation of the actual relationship which is being considered.

The concepts on the right-hand side of the diagram are conceptual. They represent the types of things that can be traced ('Traceable Type') which can be a 'Viewpoint' or a 'Viewpoint Element', together with the type of relationships that can be used ('Relationship Type'). The concepts on the left-hand side of the diagram are concrete. They represent the actual things traced ('Traceable Element') which can be a 'View' or a 'View Element', together with the type of traceability connecting them ('Traceability Relationship').

Many people consider and define traceability, fewer fully consider the way in which traceability is defined. Often where it is defined it is considered as a database schema, which although useful may not provide a direct relationship to the way a project is being conducted or what that project is trying to achieve though the provision of traceability.

## 5.3    Viewpoints

This section describes the Viewpoints that make up the Traceability Pattern. It begins with an overview of the Viewpoints, defines Rules that apply to the Pattern and then defines each Viewpoint.

### 5.3.1    Overview

The Traceability Pattern defines a number of Viewpoints as shown in the Viewpoint Relationship View (VRV) in Figure 5.3.

The Traceability Pattern defines four Viewpoints to enable the definition and capture of traceability:

- The 'Relationship Identification Viewpoint' is used to define the types of permissible Relationship Types.
- The 'Traceability Identification Viewpoint' is used to define the items that can be traced (the Traceable Types) and the types of trace that can be used between items (the Relationship Types between pairs of Traceable Types).

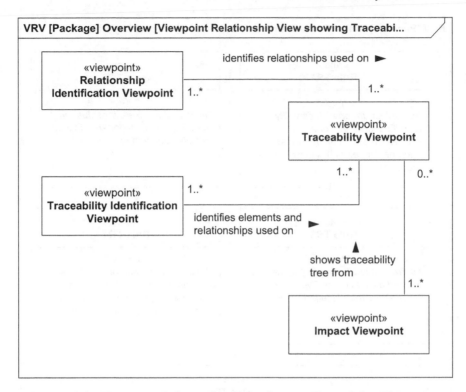

*Figure 5.3   Viewpoint Relationship View showing Traceability Viewpoints*

- The 'Traceability Viewpoint' is used to capture and visualise traceability between Traceable elements through Traceability Relationships. This may be in the form of a diagram, table, matrix etc.
- The 'Impact Viewpoint' is used to show a traceability tree for a selected Traceable Element from a Traceability View, allowing the items potentially impacted by changes to the root of the tree (the selected Traceable Element) to be identified.

### 5.3.2   Rules

Five rules apply to the four Traceability Viewpoints, as shown in the Rules Definition View (RDV) in Figure 5.4.

These five Rules are:

- 'Rule TR1: As a minimum one Traceability Identification View, one Relationship Identification View and one Traceability View must be produced.' This Rule enforces a minimum set of Views that must be produce. If this minimum set is not produced then an incomplete definition of Traceability will created.
- 'Rule TR2: All Traceability Relationships must be defined as Relationship Types on a Relationship Identification View.' This Rule is there to ensure that

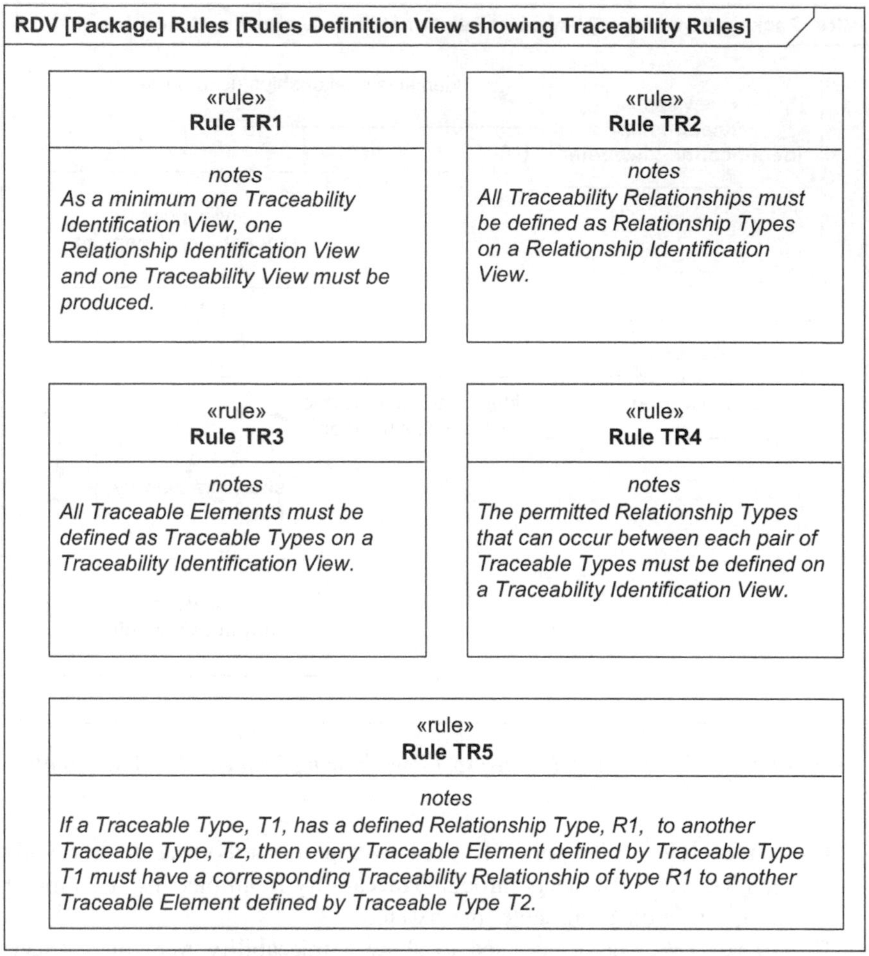

RDV [Package] Rules [Rules Definition View showing Traceability Rules]

«rule»
**Rule TR1**

*notes*
*As a minimum one Traceability Identification View, one Relationship Identification View and one Traceability View must be produced.*

«rule»
**Rule TR2**

*notes*
*All Traceability Relationships must be defined as Relationship Types on a Relationship Identification View.*

«rule»
**Rule TR3**

*notes*
*All Traceable Elements must be defined as Traceable Types on a Traceability Identification View.*

«rule»
**Rule TR4**

*notes*
*The permitted Relationship Types that can occur between each pair of Traceable Types must be defined on a Traceability Identification View.*

«rule»
**Rule TR5**

*notes*
*If a Traceable Type, T1, has a defined Relationship Type, R1, to another Traceable Type, T2, then every Traceable Element defined by Traceable Type T1 must have a corresponding Traceability Relationship of type R1 to another Traceable Element defined by Traceable Type T2.*

*Figure 5.4    Rules Definition View showing Traceability Rules*

the types of traceability relationships that will be used are defined. No Relationship Type can be used that has not been defined.

- 'Rule TR3: All Traceable Elements must be defined as Traceable Types on a Traceability Identification View.' Again, a Rule to ensure consistency. Traceability can only be established between the kinds of Traceable Types that have been defined.
- 'Rule TR4: The permitted Relationship Types that can occur between each pair of Traceable Types must be defined on a Traceability Identification View.' This Rule works with Rules TR2 and TR3 to ensure that the allowed Relationship Types that can be used between Traceable Types are defined before use.

- 'Rule TR5: If a Traceable Type, T1, has a defined Relationship Type, R1, to another Traceable Type, T2, then every Traceable Element defined by Traceable Type T1 must have a corresponding Traceability Relationship of type R1 to another Traceable Element defined by Traceable Type T2.' This wordy Rule simply says that if you have said that a particular type of traceability is to be used between particular types of elements, then when performing traceability all elements of those types must be so traced.

Note that the five Rules shown in Figure 5.4 are the minimum that are needed. Others could be added if required.

### 5.3.3 Relationship Identification Viewpoint (RIVp)

The aims of the Relationship Identification Viewpoint are shown in the Viewpoint Context View in Figure 5.5.

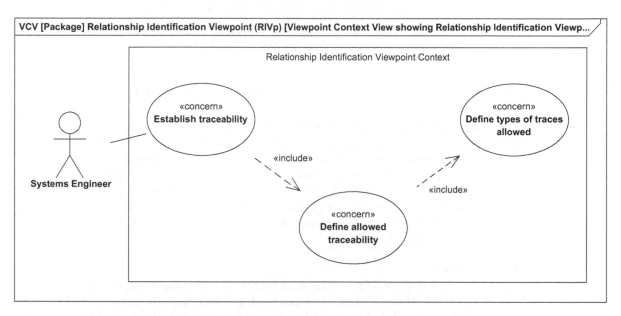

*Figure 5.5 Viewpoint Context View showing Relationship Identification Viewpoint aims*

The main aim of the Relationship Identification Viewpoint is to 'Establish traceability' through the aim of 'Define allowed traceability'. In particular, its key aim is to 'Define types of traces allowed', that is the Relationship Identification Viewpoint identifies the allowed Relationship Types that can be used to establish traceability.

#### 5.3.3.1 Description

The Viewpoint Definition View (VDV) in Figure 5.6 shows the Ontology Elements that appear on a Relationship Identification Viewpoint.

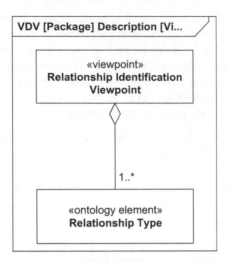

*Figure 5.6    Viewpoint Definition View showing the Ontology Elements that appear
on the Relationship Identification Viewpoint (RIVp)*

The Relationship Identification Viewpoint shows a number of Relationship Types.

Relationship Type defines the relationships that will be used to capture traceability. In many cases, traceability relationships are assumed to be loose relationships added by a requirements or traceability tool. However, this Viewpoint also enables other concepts such as parent/child relationships to represent traceability.

A Relationship Identification Viewpoint would typically be based on the Ontology of a specified Pattern or application (such as requirements engineering), adding detail relating to the types of traceability relationship to fully define the information to be captured.

### 5.3.3.2   Example

An example View that conforms to the Relationship Identification Viewpoint is shown in Figure 5.7. This, and subsequent examples, are taken from the domain of MBSE (and Model-Based Requirements Engineering in particular). However, this is only one possible application of traceability and, while perhaps the most common use within the systems engineering domain and hence the reason it was chosen for the examples, it is not the only one.

The Relationship Identification View in Figure 5.7, realised as a SysML *block definition diagram*, defines nine Relationship Types: 'Trace', 'Refinement', 'Derivation' etc., that have been identified as being the kinds of traceability needed when carrying out model-based requirements engineering.

Each of the defined Relationship Types should also be accompanied by a description of its intended use. This has been done in the following table, which also serves as a text-based version of a Relationship Identification View:

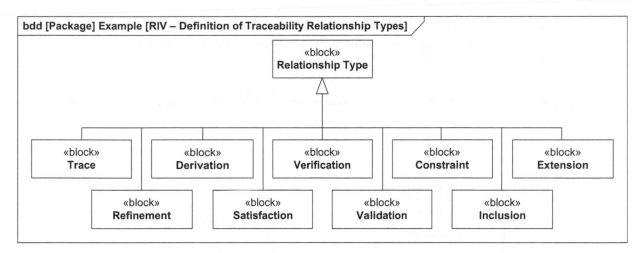

*Figure 5.7    RIV – Definition of Traceability Relationship Types*

*Table 5.1    An example of a text-based Traceability Identification View for MBRE*

| **Relationship Identification View** | |
|---|---|
| **Relationship Type** | **Description** |
| Trace | A general purpose relationship type that can be used if none of the other, more specific, relationship types is suitable. For example, to indicate that a Requirement traces to a Source Element. |
| Refinement | Indicates that one Requirement refines another or that a Use Case refines a Requirement. |
| Derivation | Indicates that one Requirement is derived, in whole or part, from another Requirement. |
| Satisfaction | Indicates that a Requirement satisfies a Rationale or Capability or that a Capability satisfies a Goal or that a System Element satisfies a Use Case. |
| Verification | Indicates that a Test Case verifies a Requirement. |
| Validation | Indicates that a Test Case validates a Use Case. |
| Constraint | Indicates that one Requirement constrains another or that one Use Case constrains another. |
| Inclusion | Indicates that one Requirement includes another as a sub-Requirement or that one Use Case includes another. |
| Extension | Indicates that one Use Case extends another under stated circumstances. |

This View, defined in Figure 5.7 and Table 5.1, fulfils Rule TR2. With the Traceability Identification View in Figure 5.10 and Table 5.2 and the Traceability View in Figure 5.13, it fulfils Rule TR1.

### 5.3.4   *Traceability Identification Viewpoint (TIVp)*

The aims of the Traceability Identification Viewpoint are shown in the Viewpoint Context View in Figure 5.8.

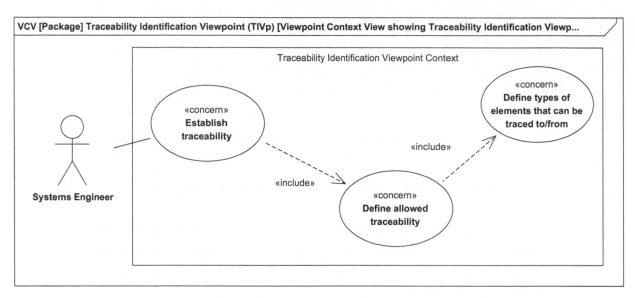

*Figure 5.8   Viewpoint Context View showing Traceability Identification Viewpoint aims*

The main aim of the Traceability Identification Viewpoint is to 'Establish traceability' through the aim of 'Define allowed traceability'. In particular, its key aim is to 'Define types of elements that can be traced to/from', that is the Traceability Identification Viewpoint identifies the allowed Traceable Types as well as the allowed Relationship Types that can be used to establish traceability between any two Traceable Types.

#### 5.3.4.1   Description

The VDV in Figure 5.9 shows the Ontology Elements that appear on the Traceability Identification Viewpoint.

The Traceability Identification Viewpoint identifies the types of elements that can be involved in traceability relationships, along with the types of trace that can be used between them.

The Traceability Identification Viewpoint shows a number of Traceable Types and Relationship Types.

Traceable Types are defined as Viewpoints and Viewpoint Elements. These define the model Viewpoints and Viewpoint Elements that can be the source and targets of traceability. If a type of Viewpoint or Viewpoint Element is not identified

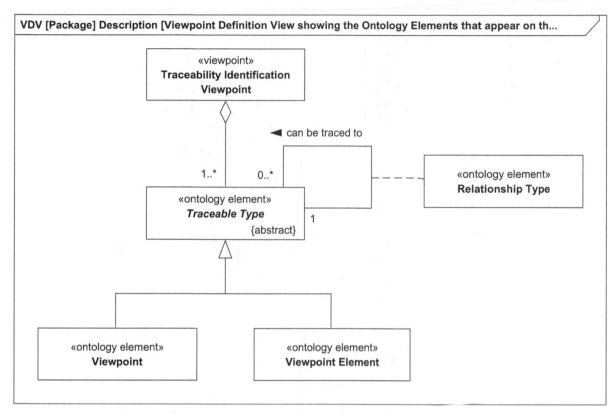

*Figure 5.9    Viewpoint Definition View showing the Ontology Elements that appear
on the Traceability Identification Viewpoint (RIVp)*

on a Traceability Identification View then no trace relationship will be able to be defined from or to that Viewpoint or Viewpoint Element.

The Relationship Types (defined on the Relationship Identification Viewpoint) are used on the Traceability Identification Viewpoint to define the allowed relationships between pairs of Traceable Types. Only those Relationship Types defined between a pair of Traceable Types can be used between them.

The Traceability Identification Viewpoint will be based on the Ontology of a specified Pattern or application (such as requirements engineering), adding detail relating to the types of items that are involved in traceability, along with the nature and direction of the traceability relationships between them to fully define the information to be captured.

### 5.3.4.2   Example

An example View that conforms to the Traceability Identification Viewpoint is shown in Figure 5.10.

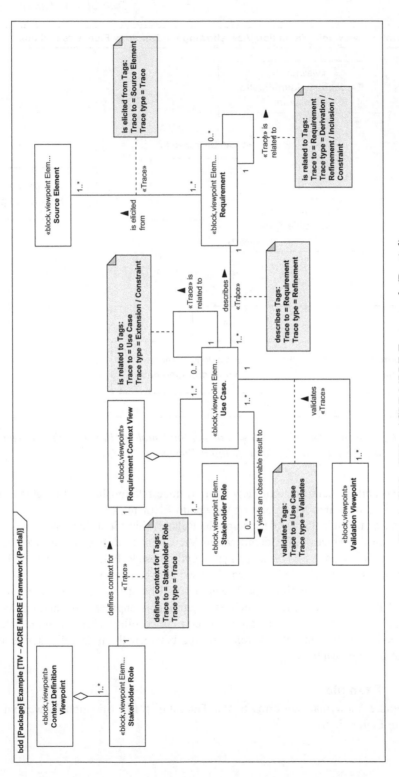

*Figure 5.10   TIV – ACRE MBRE Framework (Partial)*

This SysML *block definition diagram* shows an extract from a Model-Based Requirements Engineering framework. In this diagram there are three Viewpoints defined, the 'Context Definition Viewpoint', 'Requirement Context Viewpoint' and the 'Validation Viewpoint'. These are marked as such with the «Viewpoint» *stereotype*. A number of Viewpoint Elements are also shown, again indicated as such through the use of *stereotypes*. Note that for the Viewpoint Elements 'Requirement' and 'Source Element', their owning Viewpoints are not shown (simply to reduce the complexity of this example diagram).

The *blocks* marked as representing Viewpoints and Viewpoint Elements indicate those elements which can be traced. The *associations* with the «Trace» *stereotype* indicate the possible traceability Relationship Types that are valid between the *blocks* linked by the *association*. Where the «Trace» *stereotype* has been applied, further information about the detail of the trace is also shown through the use of *tags*:

- 'Trace to' – the direction of the trace
- 'Trace type' – the type of trace, e.g. Trace, Validate, Inclusion etc.

One item of note on the diagram concerns the *association* between the 'Stakeholder Role' (from the 'Context Definition Viewpoint') and the 'Requirement Context Viewpoint'. The trace indicated on this *association* is shown as applying in the opposite direction to the reading direction of the *association*. The *association* is read from 'Stakeholder Role' to 'Requirement Context Viewpoint' but the trace goes to the 'Stakeholder Role' rather than from it. This is due to the nature of traceability, which is often considered as going back towards the source of the information. The intention here is to show that a 'Requirement Context Viewpoint' can be traced back to its source 'Stakeholder Role' on a 'Context Definition Viewpoint'.

The same kind of information as shown in Figure 5.10 may also be shown textually as in Table 5.2, which shows a superset of the information found in Figure 5.10.

The Traceability Identification View shown in Table 5.2 defines the Traceable Types (Viewpoints and Viewpoint Elements) and Relationship Types that can exist between them for the Approach to Context-Based Requirements Engineering (ACRE) model-based requirements engineering Framework discussed in Chapter 13.

The first two columns define the Traceable Types, giving the source and destination of possible relationships. The final column defines the possible relationships that can hold between each pair of Traceable Types. Only those Relationship Types defined on a Relationship Identification View may appear in this column.

This View, defined in Figure 5.10 and Table 5.2, fulfils Rules TR3 and TR4. With the Relationship Identification View in Figure 5.7 and Table 5.1 and the Traceability View in Figure 5.13, it fulfils Rule TR1.

### 5.3.5   Traceability Viewpoint (TVp)

The aims of the Traceability Viewpoint are shown in the Viewpoint Context View in Figure 5.11.

*Table 5.2    An example of a text-based Traceability Identification View for MBRE*

| Traceability Identification View | | |
|---|---|---|
| **Traceable Type (Viewpoint or Viewpoint Element)** | | **Relationship Type** |
| **From** | **To** | |
| Requirement | Source Element | Trace |
| | Requirement | Derivation<br>Inclusion<br>Constraint<br>Refinement |
| | Rationale | Satisfaction |
| | Capability | Satisfaction |
| Use Case | Requirement | Refinement |
| | Use Case | Constraint<br>Inclusion<br>Extension |
| System Element | Use Case | Satisfaction |
| Test Case (Validation View) | Use Case | Validation |
| Test Case | Requirement | Verification |
| Requirement Context View | Stakeholder Role (on Context<br>Definition Viewpoint) | Trace |
| Capability | Goal | Satisfaction |

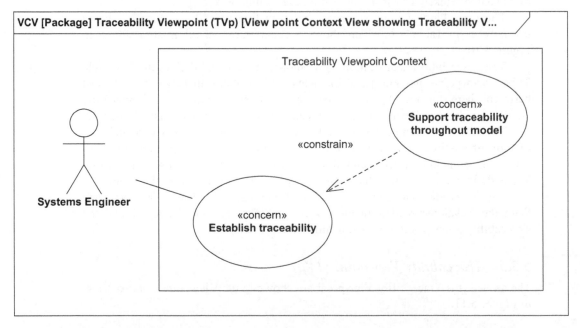

*Figure 5.11    Viewpoint Context View showing Traceability Viewpoint aims*

The main aim of the Traceability Viewpoint is to 'Establish traceability' in a model, with the constraint of 'Support traceability throughout model'. That is, the Traceability Viewpoint allows traceability to be captured and visualised throughout the model of a System; it is not something that is restricted to Requirements traceability. Traceability Views that conform to the Traceability Viewpoint may be in the form of diagrams, tables, matrices, trees, etc.

### 5.3.5.1  Description

The VDV in Figure 5.12 shows the Ontology Elements that appear on a Traceability Viewpoint.

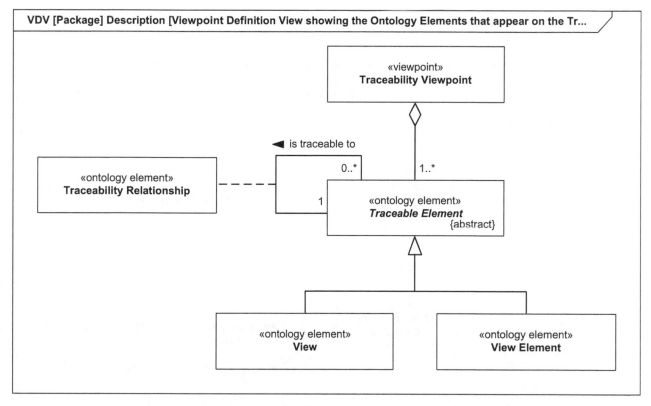

*Figure 5.12    Viewpoint Definition View showing the Ontology Elements that appear on the Traceability Viewpoint (TVp)*

The Traceability Viewpoint shows a number of Traceable Elements and the Traceability Relationships between them. Traceable Elements may be Views or View Elements and Traceability Relationships may be made View-to-View, View Element-to-View Element, View Element-to-View or View-to-View Element.

The allowed Traceability Relationships are those of the Relationship Types that are defined on a Relationship Identification View. The allowed Traceable Elements are those of the Traceable Types that are defined on a Traceability Identification View. Only the connections between pairs of Traceable Types defined on a Traceability Identification View are permitted on a Traceability View.

### 5.3.5.2   Example

An example View that conforms to the Traceability Viewpoint is shown in Figure 5.13.

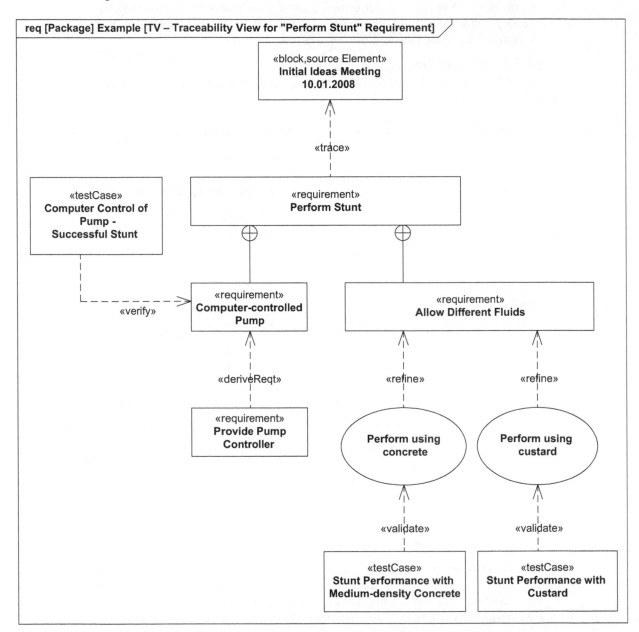

*Figure 5.13   TV Traceability View for "Perform Stunt" Requirement*

The Traceability View in Figure 5.13, here realised as a SysML *requirements diagram*, shows six different types of Traceability Relationship between a number of Traceable Elements:

- A Trace relationship between the Requirement 'Perform Stunt' and its Source Element 'Initial Ideas Meeting 10.01.2008', realised using a SysML *«trace» relationship*.
- Two Inclusion relationships between the sub-Requirements 'Computer-controlled Pump' and 'Allow Different Fluids' and their parent Requirement 'Perform Stunt', realised using SysML *nesting relationships*.
- A Derivation relationship between the Requirement 'Provide Pump Controller' and the Requirement 'Computer-controlled Pump' that it is derived from, realised using a SysML *«deriveReqt» relationship*.
- A Verification relationship between a Test Case 'Computer Control of Pump – Successful Stunt' and the Requirement 'Computer-controlled Pump' that it verifies, realised using a SysML *«verify» relationship*.
- Two Refinement relationships between the Use Cases 'Perform using concrete' and 'Perform using custard' and the Requirement 'Allow Different Fluids' that they refine, realised using a SysML *«refine» relationship*.
- Two Validation relationships between the Test Cases 'Stunt Performance with Medium-density Concrete' and 'Stunt Performance with Custard' and the Use Cases 'Perform using concrete' and 'Perform using custard' that they validate, realised using a SysML *dependency* stereotyped «validate». Note that SysML does not have a "built-in" «validate» *relationship*.

As an alternative to a graphical representation, the same information could be represented as a simple matrix as shown in Table 5.3.

Note that the Traceability View defined in Figure 5.13 and Table 5.3 has been created to show the traceability, both forwards and backwards, for a single Requirement, 'Perform Stunt'. However, this is only one possibility. There is no restriction on the focus that each Traceability View has nor on the number of levels shown. A Traceability View could be created showing, for example, a number of Source Elements and all the Requirements that trace to them (at a single level of traceability), or could concentrate on a single Source Element and everything that ultimately traces to this (multi-level traceability); this would be similar to Figure 5.13 but would show multiple Requirements tracing to the Source Element.

This View, defined in Figure 5.13 and Table 5.3, along with the Relationship Identification View in Figure 5.7 and Table 5.1 and the Traceability Identification View in Figure 5.10 and Table 5.2, fulfils Rules TR1, TR2, TR3 and TR4.

Table 5.3 An example of a Traceability View represented as a matrix

**Traces From**

| Traces To ↓ | «requirement» Perform Stunt | «requirement» Computer-controlled Pump | «requirement» Allow Different Fluids | «requirement» Provide Pump Controller | «testCase» Computer Control of Pump – Successful Stunt | «Use Case» Perform using concrete | «Use Case» Perform using custard | «testCase» Stunt Performance with Medium-density Concrete | «testCase» Stunt Performance with Custard |
|---|---|---|---|---|---|---|---|---|---|
| «Source Element» Initial Ideas Meeting 10.01.2008 | Trace «trace» | | | | | | | | |
| «requirement» Perform Stunt | | Inclusion (nesting) | | | | | | | |
| «requirement» Computer-controlled Pump | | | Inclusion (nesting) | Derivation «deriveReqt» | | | | | |
| «requirement» Allow Different Fluids | | | | | Verification «verify» | Refinement «refine» | Refinement «refine» | | |
| «Use Case» Perform using concrete | | | | | | | | Validation «validate» | |
| «Use Case» Perform using custard | | | | | | | | | Validation «validate» |

### 5.3.6 *Impact Viewpoint (IVp)*

The aims of the Impact Viewpoint are shown in the Viewpoint Context View in Figure 5.14.

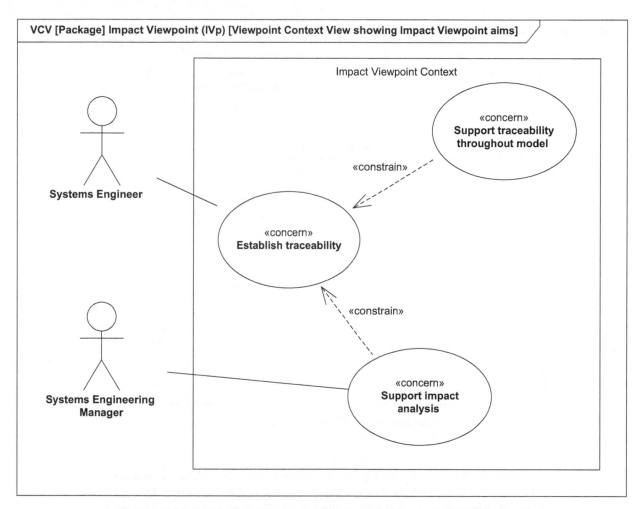

*Figure 5.14 Viewpoint Context View showing Impact Viewpoint aims*

The main aim of the Impact Viewpoint is to 'Establish traceability' in a model, with the constraints of 'Support traceability throughout model' and 'Support impact analysis'. An Impact View that conforms to the Impact Viewpoint is used to show a traceability tree for a selected Traceable Element from a Traceability View, allowing the items (the other Traceable Elements connected to it through a network of Traceability Relationships) that may be impacted by changes to the root of the tree (the selected Traceable Element) to be identified.

### 5.3.6.1 Description

The VDV in Figure 5.15 shows the Ontology Elements that appear on an Impact Viewpoint.

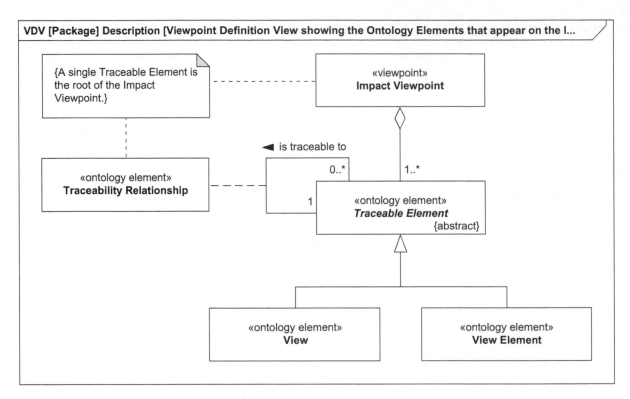

*Figure 5.15 Viewpoint Definition View showing the Ontology Elements that appear on the Impact Viewpoint (IVp)*

Like the Traceability Viewpoint, the Impact Viewpoint shows a number of Traceable Elements and the Traceability Relationships between them. Indeed, the Ontology Elements that can appear on the Impact Viewpoint and the rules governing consistency of those elements are the same as for the Traceability Viewpoint.

Where the two Views differ is in the form and intent of the Views created based on the Viewpoints. A Traceability View can show any number of levels of traceability and may or may not have a single "root" element that everything traces to or from. An Impact View, however, always has a single root element and typically shows all the levels of traceability starting from that root element. Whereas a Traceability View may show a web of related elements, an Impact View always shows a tree of related elements, starting from the element that is the root of the tree.

### 5.3.6.2    Example

An example View that conforms to the Impact Viewpoint is shown in Figure 5.16.

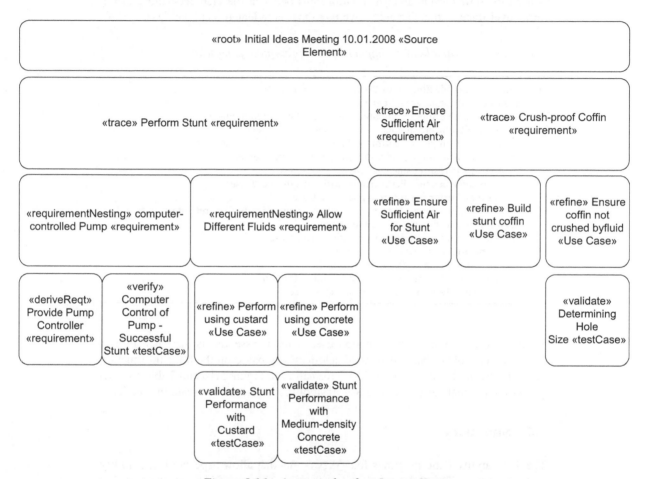

*Figure 5.16    An example of an Impact View*

The Impact View in Figure 5.16 shows the forward impact tree for the Source Element 'Initial Ideas Meeting 10.01.2008', showing all 15 potential model elements that may be impacted by a change to the Source Element. Note that there is an implicit meaning in the layout of the diagram: items in a given level trace to the item lying over them in the level above. The nature of the trace relationship is shown in the *stereotype* that forms the first part of the name of each item and the type of item traced is shown in the *stereotype* that forms the last part of the name. For example, from the diagram it can be seen that the Use Cases 'Build stunt coffin' and 'Ensure coffin not crushed by fluid' both trace, through refinement, to the Requirement 'Crush-proof Coffin'.

Other graphical representations are, of course, possible. The same information may also be presented as text, as shown in Table 5.4. Here, the level of indentation is used to show the hierarchical relationships between the elements; those at the same level trace to the element above them that is indented one level less.

*Table 5.4　An example of an Impact View represented using text*

«root» Initial Ideas Meeting 10.01.2008 «Source Element»
  «trace» Perform Stunt «requirement»
    «requirementNesting» Computer-controlled Pump «requirement»
      «deriveReqt» Provide Pump Controller «requirement»
      «verify» Computer Control of Pump - Successful Stunt «testCase»
    «requirementNesting» Allow Different Fluids «requirement»
      «refine» Perform using custard　　「«Use Case»
        «validate» Stunt Performance with Custard «testCase»
      «refine» Perform using concrete «Use Case»
        «validate» Stunt Performance with Medium-density Concrete «testCase»
  «trace» Ensure Sufficient Air «requirement»
    «refine» Ensure Sufficient Air for Stunt «Use Case»
  «trace» Crush-proof Coffin «requirement»
    «refine» Build stunt coffin «Use Case»
    «refine» Ensure coffin not crushed by fluid «Use Case»
      «validate» Determining Hole Size «testCase»

Whatever the graphical representation used, what is essential is that such Impact Views are capable of being generated automatically based on the trace information added to the model of a System. The information in Figure 5.16 and Table 5.4 was generated automatically from the SysML tool that was used to create the model.

## 5.4　Summary

The Traceability Pattern defines four Viewpoints that allow aspects of traceability to be captured. The Relationship Identification Viewpoint defines the possible types of traceability relationships that may be used. The Traceability Identification Viewpoint defines the types of model elements that can be the sources and targets of traceability and the relationships (from the Relationship Identification Viewpoint) that can hold between them. The Traceability Viewpoint shows the *actual* traceability relationships that hold between model elements that conform to the allowed types for traceability (as defined on the Traceability Identification Viewpoint). Finally, the Impact Viewpoint allows a traceability *impact* tree to be produced for a selected *root* model element, aiding in the identification of those model elements that may be impacted by changes to the selected root element.

## 5.5    Related Patterns

If using the Traceability Pattern, the following Patterns may also be of use:

- Testing
- Interface Definition.

## Reference

Holt, J. and Perry, S. *SysML for Systems Engineering; 2nd Edition: A model-based approach*. London: IET Publishing; 2013.

## Further readings

Object Management Group. *SysML website* [online]. Available at: http://www. omgsysml.org [Accessed February 2015].

Holt, J. *UML for Systems Engineering – Watching the Wheels*. 2nd Edition. London: IEE Publishing; 2004 (reprinted IET Publishing; 2007).

Rumbaugh, J., Jacobson, I. and Booch, G. *The Unified Modeling Language Reference Manual*. 2nd Edition. Boston, MA: Addison-Wesley; 2005.

Stevens, R., Brook, P., Jackson, K. and Arnold, S. *Systems Engineering – Coping with Complexity*. London: Prentice Hall Europe; 1998.

Boehm, B. W. 'Verifying and Validating Software Requirements and Design Specifications,' reprinted in Boehm, B. W. (ed.), *Software Risk Management*, pp. 205–218, IEEE Computer Society Press, 1989.

*Chapter 6*

# Test Pattern

## 6.1 Introduction

Testing is an essential part of any system development. Testing must demonstrate two main aims of the development by answering two questions: [Boehm]

- Have we built the right system? (validation)
- Have we built the system right? (verification)

Both verification and validation (V&V) can take on many different forms so it is important that the system can be tested in many different ways, ideally using a single model as a basis for all the testing activities. In the context of safety-critical systems development it is vital to justify that adequate testing activities have been performed, and sufficient test cases have been exercised on the system under test.

As a consequence it is desirable to adopt an organised approach to define test schedules that define the creation and execution of test cases.

Such an approach must be produced in a systematic way. If test schedules can exploit tool support automating some of the activities involved, then it is essential that these tools conform to such an organised approach, because otherwise unnecessary effort will have to be spent on mapping the artefacts produced by the tools to the evidence required to justify adequateness and comprehensiveness of the test schedule. This model-based approach forms an essential basis for model-based testing.

### 6.1.1 Pattern aims

The Test Pattern defines a structure of the model that can then be used as a basis for model-based testing. The main aims of the Test Pattern are shown in Figure 6.1.

The Context for the Testing Pattern is shown in Figure 6.1. The key aim of the Pattern is to 'Define tests'. This has two main inclusions, which are:

- 'Define testing context'. This is an essential part of the Pattern as it defines exactly what is expected from the testing, along with any external factors that need to be taken into account, such as standards, test equipment, data, etc.
- 'Define test set up'. This allows the Test Schedule and Test Sets to be defined.
- 'Define test cases'. This allows the actual tests that are to be executed to be defined.

Notice that the Test Context is so important that it constrains the other two use cases on this diagram.

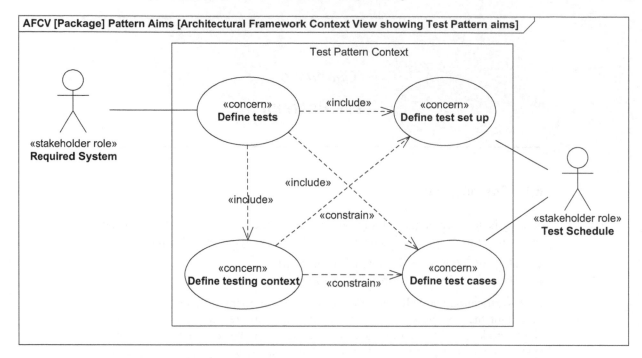

*Figure 6.1    Architectural Framework Context View showing Testing Pattern Context Viewpoint aims*

## 6.2    Concepts

The main concepts covered by the Testing Pattern are shown in the Ontology Definition View (ODV) in Figure 6.2.

Figure 6.2 shows the Testing Pattern ontology by identifying the main concepts and the relationships between them.

A fundamental part of the 'Test Case' Pattern is, quite naturally, the 'Test Case'. A 'Test Case' is executed against a 'Testable Element' using 'Test Data'. One or more 'Test Case' is collected into a 'Test Set', one or more of which make up a 'Test Schedule'. It may be possible than more than one Test Set may be required depending on the Testing Context. For example, one Test Set may be required for validation, one for verification, etc. However, it is not good enough to simply define a structure for a 'Test Case' and then realise it, as we need to understand why the testing is necessary in the first place, what are the constraints, what other systems are required, what is the level of testing required and so on. This overall rationale behind the tests is represented by the 'Testing Context'. The 'Testing Context' comprises three main elements:

- One or more 'Testing Need'. This represents the needs of the testing that will be carried out (for example: verification, validation, etc.) and will also define

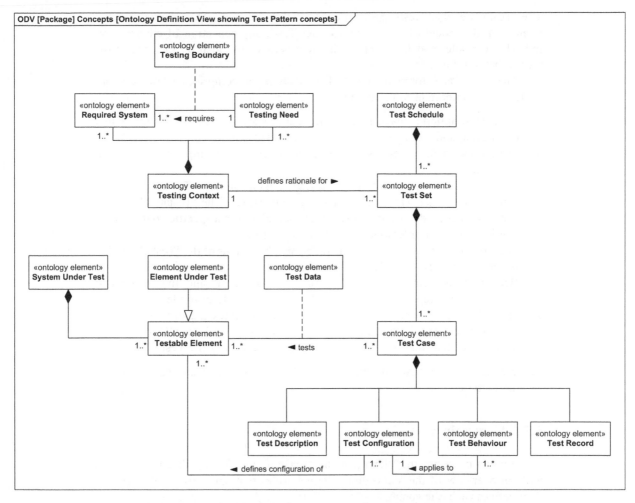

*Figure 6.2   Ontology Definition View showing Testing concepts*

any constraints on the testing, such as the scope of the tests, any specific techniques, and so on. It should be noted that the 'Testing Need' does *not* represent the needs of the system or any of the other stakeholders.

- One or more 'Required System'. This represents any system that falls outside the testing boundary that may be required for the successful execution of the tests. Examples of such required systems include the environment, standards, test data, etc.

- 'Testing Boundary'. The 'Testing Boundary' represents the partition between what is included on the model (that satisfies the 'Testing Need') to be used for the testing and everything that is not part of the model, but is necessary to perform the testing (the 'Required System').

The 'Test Case' itself tests one or more 'Testable Element'. This can be any element from the model of the 'System Under Test', any collection of elements or, indeed, the whole model. The main aim here is to simply identify the area of the model that will be tested.

The 'Test Case' forms the heart of the Pattern and comprises a few key concepts that are important to understand:

- 'Test Description', which is quite straightforward and simply defines a text description and a unique identifier for the 'Test Case'.
- 'Test Data', that represents any data that is used as an input to the testing activities.
- 'Test Configuration' defines the configuration of one or more 'Testable Element' that is to be tested. This is a structural representation of the 'Testable Element' and any other model elements that are connected to it in a specific configuration that is required in order to satisfy a 'Testing Need'.
- 'Test Behaviour' that describes an anticipated behaviour of the 'Test Case' that is applied to a 'Test Configuration'. There will usually be more than one 'Test Behaviour' associated with each 'Test Configuration' and the anticipated behaviour may represent both desirable and undesirable examples.
- 'Test Record'. Whereas the 'Test Behaviour' describes the test that will be performed by defining the expected behaviour, the 'Test Record' captures the actual behaviour that is observed during the execution of the Test Case.

These concepts are visualised through a number of Views that are together in the Testing Pattern.

## 6.3   Viewpoints

This section describes the Viewpoints that make up the Testing Pattern. It begins with an overview of the Viewpoints, defines Rules that apply to the Pattern and then defines each Viewpoint.

### 6.3.1   OverView

The Testing Pattern defines a number of Viewpoints as shown in the Viewpoint Relationship View (VRV) in Figure 6.3.

Figure 6.3 shows that the Testing Pattern has three main Viewpoints:

- The 'Testing Context Viewpoint' that defines the rationale for the Viewpoint and identifies a number of needs and constraints that must be satisfied in order that the test campaign be deemed successful.
- The 'Test Set-up Viewpoint' describes the structure of the Test Schedule and Test Set (using the 'Test Structure Viewpoint') and then the execution of the Test Schedule (using the 'Test Schedule Behaviour Viewpoint') and the Test Set (using the 'Test Set Behaviour Viewpoint')

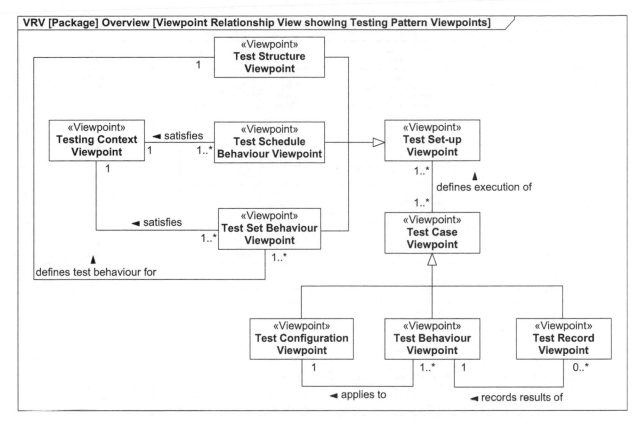

*Figure 6.3    Viewpoint Relationship View showing Testing Pattern Viewpoints*

- The 'Test Case Viewpoint' that identifies the scope of the Test Case (using the 'Test Configuration Viewpoint' that defines how the Test Case is set up), the anticipated behaviour of the Test Case (using the 'Test Behaviour Viewpoint' that shows the preferred behaviour of the Test Case) and captures the results of the Test Case (using the 'Test Record Viewpoint').

These Viewpoints are defined in the following sections.

### 6.3.2    Rules

Sixteen rules apply to the Testing Pattern Viewpoints, as shown in the Rules Definition View (RDV) in Figure 6.4.

The Testing Pattern has 16 Rules defined that are shown in Figure 6.4 and that are described as:

- 'TP1: A Testing Context must exist that defines every Testing Need.' This is fundamental to the Testing Pattern as it defines the testing Needs for the whole Test Schedule. Without this Rule, none of the others can be successful.

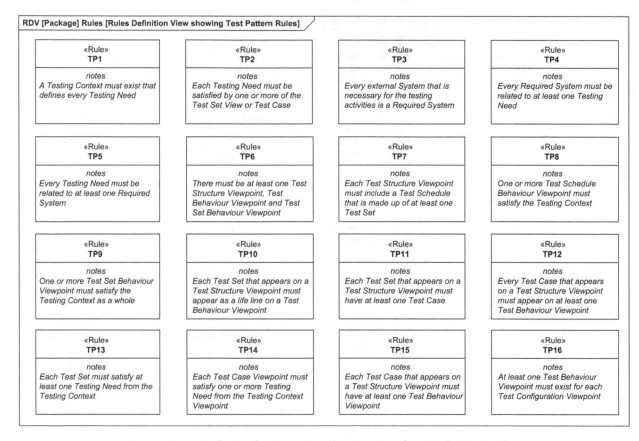

*Figure 6.4    Rules Definition View showing Interface Definition Rules*

- 'TP2: Each Testing Need must be satisfied by one or more of the Test Set View or Test Case.' This represents a coverage Rule in that all of the Testing Needs must be covered by the overall Test Schedule.
- 'TP3: Every external System that is necessary for the testing activities is a Required System.' This ensures that all external systems necessary for successful testing have been identified.
- 'TP4: Every Required System must be related to at least one Testing Need.' Any Required System that is not connected to a Testing Need is redundant and, therefore, should not feature on this View.
- 'TP5: Every Testing Need must be related to at least one Required System.' Every Testing Need must be connected to a Required System as the Required System identifies test models, external data sources, test equipment, etc. This connection may be either direct or indirect.
- 'TP6: There must be at least one Test Structure Viewpoint, Test Behaviour Viewpoint and Test Set Behaviour Viewpoint.' A View for each of the Test Set-up Viewpoints must exist, otherwise the testing activities are not defined.

- 'TP7: Each Test Structure Viewpoint must include a Test Schedule that is made up of at least one Test Set.' There must be at least one Test Schedule as it is the Test Schedules that ultimately own all of the Test Set and hence Test Cases.
- 'TP8: One or more Test Schedule Behaviour Viewpoint must satisfy the Testing Context.' The overall Test Schedule Behaviours must satisfy the complete set of Testing Needs in the Testing Context.
- 'TP9: One or more Test Set Behaviour Viewpoint must satisfy the Testing Context as a whole.' The overall Test Ste Behaviours must satisfy the Testing Needs in the Testing Context. The Rule is related to the previous Rule, TP8.
- 'TP10: Each Test Set that appears on a Test Structure Viewpoint must appear as a life line on a Test Behaviour Viewpoint.' This Rule enforces consistency between the Test Structure Views and the Test Behaviour Views by ensuring that the Test Behaviours that are executed are based on Test Sets.
- 'TP11: Each Test Set that appears on a Test Structure Viewpoint must have at least one Test Case.' Test Sets are not permitted to be empty – they must contain at least one Test Case.
- 'TP12: Every Test Case that appears on a Test Structure Viewpoint must appear on at least one Test Behaviour Viewpoint.' All Test Cases must be executed at some point during the testing activity otherwise it is redundant and should not appear.
- 'TP13: Each Test Set must satisfy at least one Testing Need from the Testing Context.' This is a lower-level coverage check that ensures that all Test Sets ultimately relate back to a Testing Need.
- 'TP14: Each Test Case Viewpoint must satisfy one or more Testing Need from the Testing Context Viewpoint.' This is a lower-level coverage check that ensures that all Test Set Case satisfies at least one Testing Need.
- 'TP15: Each Test Case that appears on a Test Structure Viewpoint must have at least one Test Behaviour Viewpoint.' All Test Cases must have their behaviour defined.
- 'TP16: At least one Test Behaviour Viewpoint must exist for each Test Configuration Viewpoint.' All Test Configurations must form the basis of some testing via the Test Behaviour Views otherwise they are redundant.

Note that the 16 Rules shown in Figure 6.4 are the minimum that are needed. Others could be added if required.

### 6.3.3 *Testing Context Viewpoint (TCVp)*

The aims of the Testing Context Viewpoint are shown in the Viewpoint Context View in Figure 6.5.

The main aim of this Viewpoint is to provide the context of the testing activity or, to be more specific, to define:

- The 'Testing Boundary' of the testing activity, visualised on a *use case diagram* using a *system boundary*.
- The 'Testing Need' of the testing activity, visualised on a *use case diagram* using *use cases*.

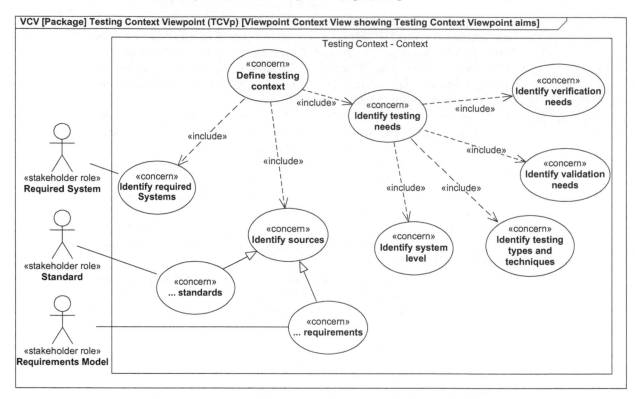

*Figure 6.5   Viewpoint Context View showing Testing Context Viewpoint aims*

- Any 'Required System' outside the boundary that is necessary to complete the testing activity. This may include a number of elements, such as the 'Requirements Model' that may be necessary for validation, one or more 'Standard' that may be required for quality-based testing, specific test data sets, any testing tools needed, etc.

This Viewpoint really drives the whole of the testing activity and will have Rules defined that can be applied to ensure consistency within the Testing Framework.

This Viewpoint can also be used to create a library of testing Use Cases and then to show how these Use Cases may be satisfied. This will be particularly useful for specific testing types, for example, equivalence class partition testing may be realised using a particular set of Views which may then be used as a basis for generating automated tests.

### 6.3.3.1   Description

The Viewpoint Definition View (VDV) in Figure 6.6 shows the Ontology Elements that appear on a Testing Context Viewpoint.

Figure 6.6 shows the Ontology for the Testing Context Viewpoint.

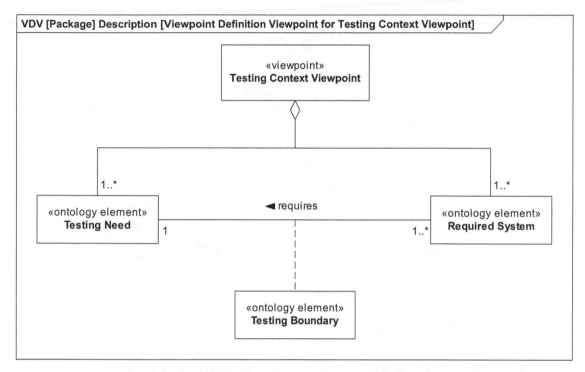

*Figure 6.6  Viewpoint Definition View showing the Ontology Elements that appear on the Testing Context Viewpoint (TCVp)*

The main two Ontology Elements in this Viewpoint are:

- 'Testing Need' that represents the fundamental need for the whole testing activity.
- 'Required System' that represents elements external to the System that are necessary for testing. Examples of Required Systems include standards, test equipment, test data, etc.

The Required System and the Testing Need are related via the Testing Boundary.

### 6.3.3.2  Example Testing Context View

An example of the Testing Context View is shown in Figure 6.7.

The 'Testing Context View' in this example is visualised in SysML using a *use case diagram*.

The main aim of this View is to provide the context of the testing activity or, to be more specific, to define:

- The 'Testing Boundary' of the testing activity, visualised on a *use case diagram* using a *system boundary*.
- The 'Testing Need' of the testing activity, visualised on a *use case diagram* using *use cases*.

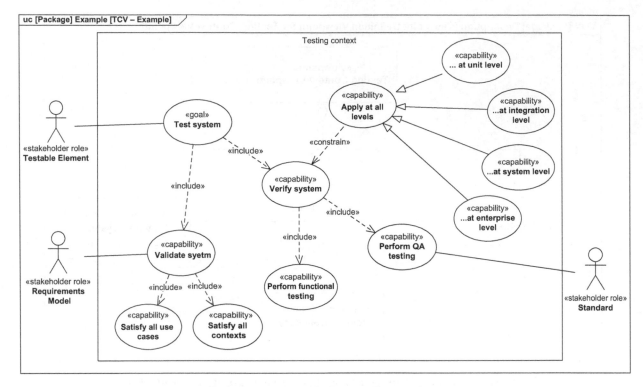

*Figure 6.7    Example Testing Context View*

- Any 'Required System' outside the boundary that is necessary to complete the testing activity. This may include a number of elements, such as the 'Requirements Model' that may be necessary for validation, one or more 'Standard' that may be required for quality-based testing, specific test data sets, any testing tools needed, etc. Each Required System is represent by an *actor* on the *use case diagram*.

On top of these elements there will also be a number of relationships between the different elements, such as:

- Relationships between the use cases, typically the standard SysML dependency-based relations (*include*, *extend*, *constrain*, etc.)
- Relationships between the *use cases* and the external *actors*, using the standard SysML *associations*.

In this example, the main aim of the Testing Context is to 'Test System' which has two inclusions:

- 'Validate system' which in this case is further specified by stating that in order to validate the system, there are two aims, which re to 'Satisfy all use cases' and to 'Satisfy all contexts'.

- 'Verify system' which in this example has two specific types of testing which are 'Perform functional testing' and to 'Perform QA testing'. This aim is further qualified by stating that the verification should 'Apply at all levels' which are to apply: '... at unit level', '... at integration level', '... at system level' and '... at enterprise level'.

This View really drives the whole of the testing activity and will have rules defined that can be applied to ensure consistency within the Testing Pattern.

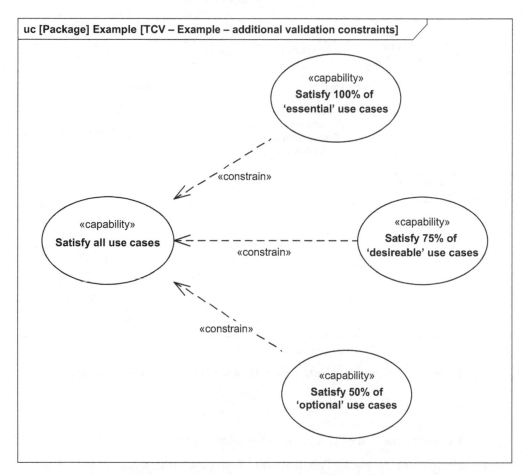

*Figure 6.8   Example Testing Context View – defining additional validation constraints*

Figure 6.8 shows how one of the *use cases* from the Testing Context View has been expanded to show more detail. In this example, the 'Satisfy all use cases' *use case* has been broken down into three lower-level *use cases* that describe more-specific testing criteria associated with the *use cases*.

### 6.3.4    Test Set-up Viewpoint (TSVp)

The aims of the Test Set-up Viewpoint are shown in the Viewpoint Context View in Figure 6.9.

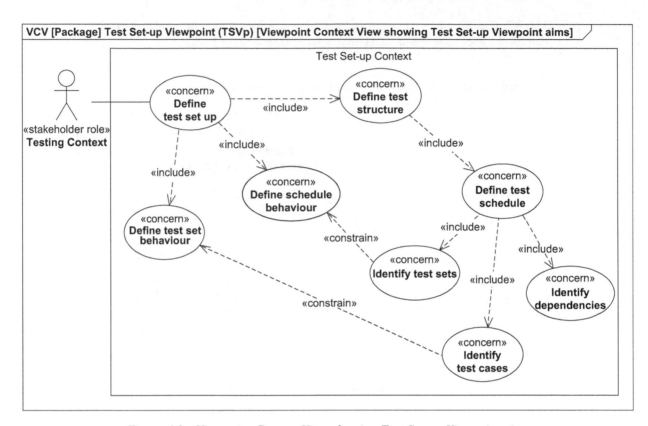

*Figure 6.9    Viewpoint Context View showing Test Set-up Viewpoint aims*

The main aims of the 'Test Set-up Viewpoint' are to:

- Identify one or more 'Test Set' and 'Test Case' that make up the 'Test Schedule'.
- To show how the 'Test Schedule' is performed by defining the order of execution of one or more 'Test Set'.
- To show how each 'Test Set' is performed by defining the order of execution of one or more 'Test Case'.

The Test Set-up Views do not describe the tests themselves but rather the execution of the Test Schedule and its associated Test Sets and Test Cases. The totality of the test set-up Views forms the basis of the test campaign.

### 6.3.4.1  Description

The VDV in Figure 6.10 shows the Ontology Elements that appear on a Test Set-up Viewpoint.

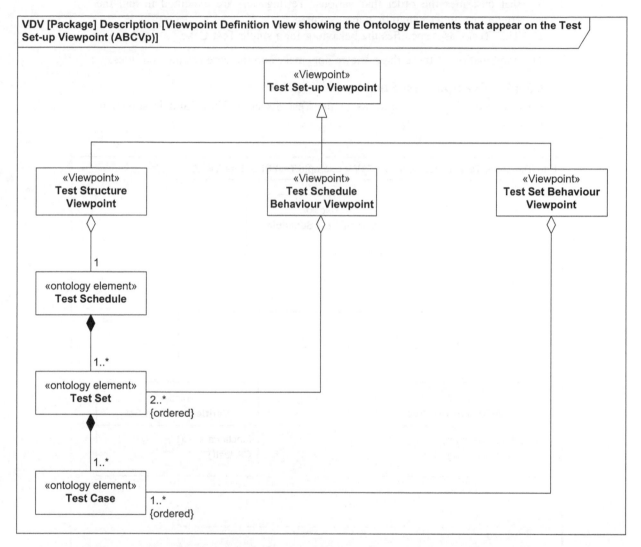

*Figure 6.10    Viewpoint Definition View showing the Ontology Elements that appear on the Test Set-up Viewpoint*

In Figure 6.10 it can be seen that this View has three types – the 'Test Structure Viewpoint', the 'Test Schedule Behaviour Viewpoint' and the 'Test Set Behaviour Viewpoint'.

- The 'Test Structure Viewpoint' represents the whole of the testing in its entirety and defines the 'Test Schedule'. Each 'Test Schedule' that is identified is further described by defining its component set of one or more 'Test Schedule'.

- The 'Test Schedule Behaviour Viewpoint' represents the highest level View that describes the order that various 'Testing Set' are executed in and the conditions under which they are executed.
- The 'Test Case' specifies the behaviour for a single Test Case.

The combination of these three Viewpoints makes up the core testing activities.

### 6.3.4.2 Example Test Structure Viewpoint

An example View that conforms to the Test Structure Viewpoint is shown in Figure 6.11.

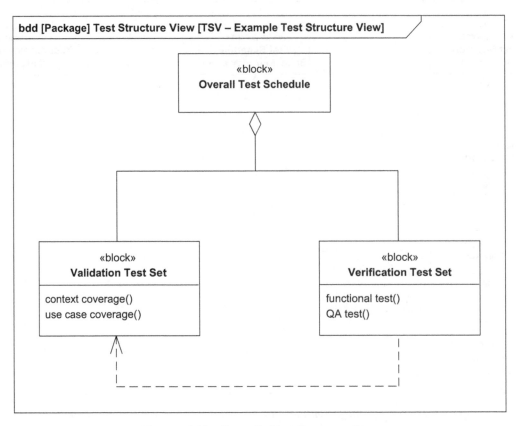

*Figure 6.11   Example Test Structure View*

The 'Test Structure View' will be visualised in SysML using a *block definition diagram* showing:

- The 'Test Schedule', represented as a SysML *block* stereotyped by «Test Schedule», and that is made up of a number of 'Test Set'.

- One or more 'Test Set', represented as SysML *blocks* that are stereotyped by «Test Set» and that are aggregated into the 'Test Schedule'.
- One or more 'Test Case', represented as SysML *operations* that are stereotyped by «Test Case».

*Dependencies* between Test Sets can also be used to show where one Test Set may depend on the successful execution of another Test Set.

The example here shows that the 'Overall Test Schedule' is made up of two Test Sets, which are:

- 'Validation Test Set' which has two Test Cases that are represented by the *operations* on the *block* and that are: 'context coverage' and 'use case coverage'.
- 'Verification Test Set' which has two Test Cases that are represented by the *operations* on the *block* and that are: 'functional test' and 'QA test'.

The *dependency* shown here represents the fact that the 'Verification Test Set' cannot be carried out without the 'Validation Test Set' having been carried out.

### 6.3.4.3   Example Test Schedule Behaviour Viewpoint

An example View that conforms to the Test Schedule Viewpoint is shown in Figure 6.12.

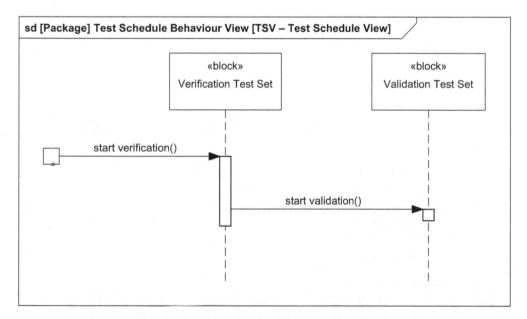

*Figure 6.12   Test Schedule Behaviour View*

Figure 6.12 shows an example of a Test Schedule Behaviour View. The diagram shows the order of execution of a number of Test Sets.

The 'Test Schedule Behaviour Viewpoint' may be represented in SysML as a *sequence diagram* where each 'Test Set' is visualised using a *life line*, and the ordering between them is shown using *messages*. In order to show the different paths of execution, then combined fragments may be necessary.

As each 'Test Schedule Behaviour Viewpoint' defines a scenario, it is possible to define a number of these Views (for example showing what happens when all test sets are passed, or when some fail).

### 6.3.4.4  Example Test Set Behaviour Viewpoint

An example View that conforms to the Test Behaviour Viewpoint is shown in Figure 6.13.

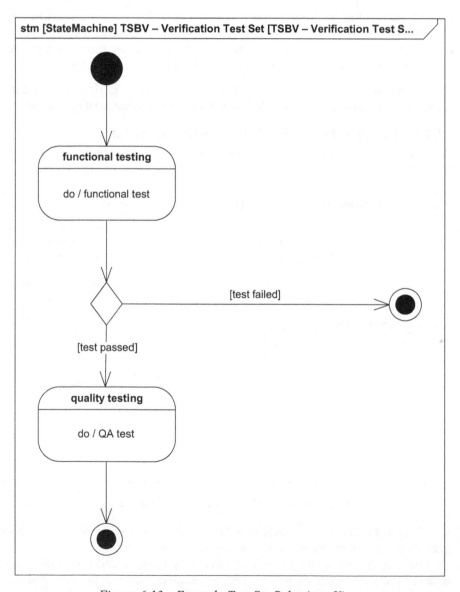

*Figure 6.13   Example Test Set Behaviour View*

Figure 6.13 shows an example of a Test Set Behaviour View. This View shows what happens inside each Test Set when it is executed in terms of the order that Test Cases are executed and any conditions that may apply.

The behaviour of each Test Set is represented in SysML using a *state machine diagram* and each Test Case is shown as an *activity* within a *state*.

The example here shows the order that the Test Cases must be executed in. In this case the 'functional test' Test Case is executed first and only upon its successful completion is the 'QA test' executed. In the event that the test fails, then the testing is halted and does not continue. This also explains the *dependency* in Figure 6.11.

## 6.3.5  *Test Case Viewpoint (TCVp)*

The aims of the Test Case Viewpoint are shown in the Viewpoint Context View in Figure 6.14.

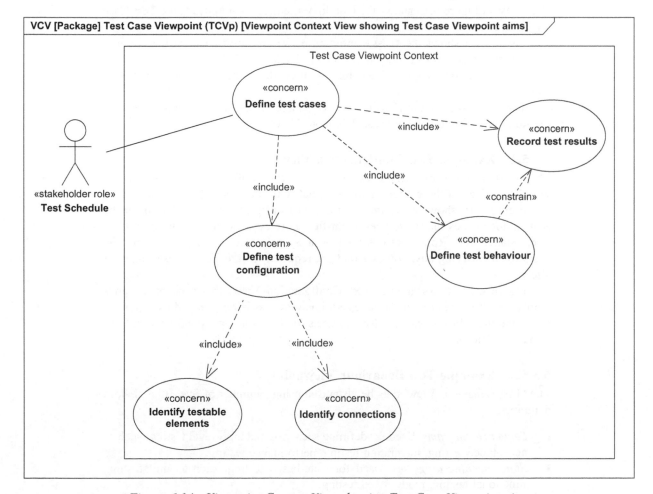

*Figure 6.14   Viewpoint Context View showing Test Case Viewpoint aims*

Figure 6.14 shows the Context for the Test Case Viewpoint.

The main aims of this Viewpoint are to 'Define test cases', which include:

- 'Define Test Configuration' that allows the Test Cases to be set up and that includes identifying the Testable Elements that will be the subject of the tests ('Identify testable elements') and the relationships between them ('Identify connections').
- 'Define test behaviour', where the behaviour for each Test Case is defined and that must exist in order for the Test Results to be captured.
- 'Record test results', where the information captured during the execution of the Test Cases is recorded.

The 'Test Schedule' is shown as an *actor* on this diagram.

### 6.3.5.1   Description

The VDV in Figure 6.15 shows the Ontology Elements that appear on a Test Case Viewpoint.

Figure 6.15 shows the 'Test Configuration Viewpoint' that identifies the part of the model that is under test – the 'Element Under Test'.

The 'Test Behaviour Viewpoint' defines the behaviours for the 'Test Configuration'.

The 'Test Record View' will not be considered in great detail here, as it uses the same visualisation as the 'Test Behaviour View'.

### 6.3.5.2   Example Test Configuration View

This section shows two examples of how a Test Configuration View may be visualised. Each of these Views may be visualised using a SysML *block diagram.*

Figure 6.16 shows an example of a Test Configuration View based on the Stakeholder Roles that have been identified as relevant to the testing activities. Each Stakeholder Role is represented by a SysML *block.* These will form the basis for identifying any Contexts that are required for the Test Case Behaviour Views.

Figure 6.17 shows another Test Configuration View, this time based on a number of Processes that must be tested. Each Process is represented by a SysML *block* and these Processes will also appear as *life lines* on corresponding Test Case Behaviour Views.

### 6.3.5.3   Example Test Behaviour Viewpoint

The 'Test Behaviour View' may be visualised using a number of SysML behaviour diagrams.

- *Sequence diagrams.* Used for defining scenarios and high-level testing, such as acceptance testing, integration testing, performance testing.
- *State machine diagrams.* Used for state-based testing, such as unit testing, functional testing, mode-based testing.

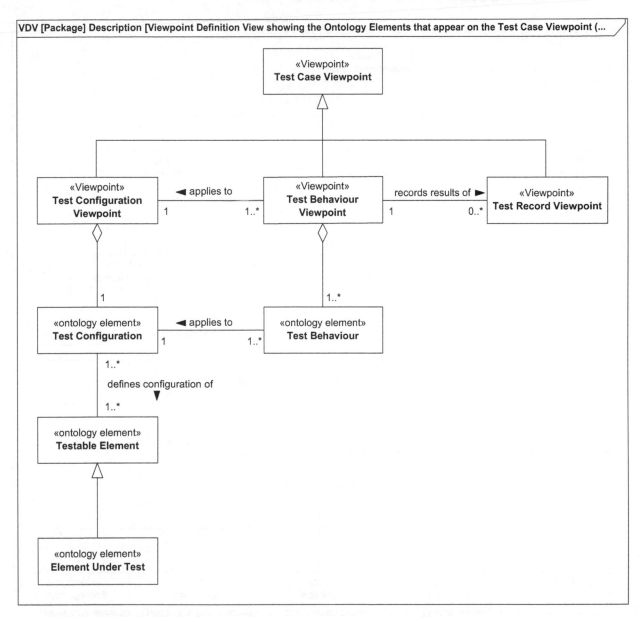

*Figure 6.15   Viewpoint Definition View showing the Ontology Elements that appear on the Test Case Viewpoint*

- *Activity diagrams.* Used for detailed testing, such as unit testing, functional; testing, process-based testing.
- *Parametric diagrams.* Used for defining mathematical-based scenarios for testing at any level.

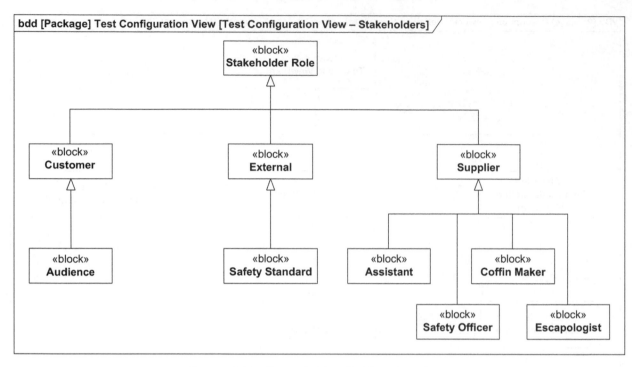

*Figure 6.16   Example Test Configuration View*

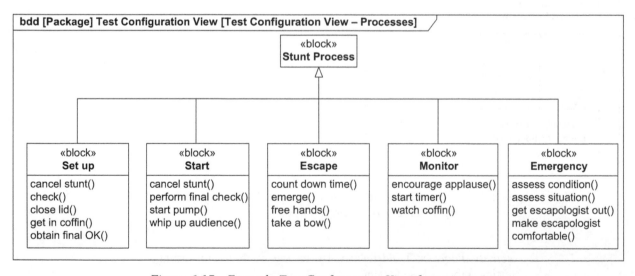

*Figure 6.17   Example Test Configuration View for processes*

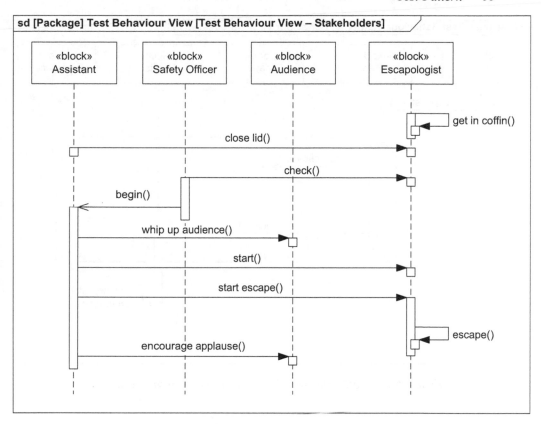

*Figure 6.18    Example Test Behaviour View*

Figure 6.18 shows an example of a Test Case Behaviour View where a *sequence diagram* is used to show the interactions between Stakeholder Roles in the execution of a specific Test Case.

In the example shown here, the Scenario focuses on the interactions between the various Stakeholder Roles and the order of these interactions. So in order to perform this Test Case, the first thing that happens is that the 'Escapologist' has a self-interaction which is to 'get in coffin'. The next thing that happens is that the 'Assistant' closes the lid, and so on.

It is possible to use this View to capture not only what constitutes a successful Test Case but also how the System must be tested to respond to some sort of problem, as shown in the next example.

Figure 6.19 shows another example Test Case Behaviour View but, this time, the Test Case reflects what must happen when a problem occurs during the execution of the System.

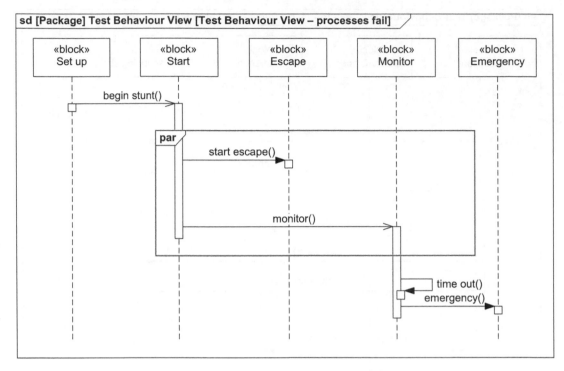

*Figure 6.19   Test Behaviour View – processes, failure scenario*

The example here shows how the Processes from Figure 6.17 are executed and, therefore, how they must be tested. In this case the 'Set up' Processes is executed and then the 'Start' Process is executed which simultaneously sends out two *messages* 'start escape' to the 'Escape' Process and 'monitor' to the 'Monitor' Process. In this example, the 'Monitor' Process detects a 'time out' and invokes the 'Emergency' Process.

### 6.3.5.4   Example Test Record Viewpoint

The Test Record View captures any test results that are generated during the execution of a Test Case.

Figure 6.20 shows an example of how test results may be captured and then annotated to a copy of the original Test Behaviour View. In this example, the Test Case was executed successfully and a simple SysML *note* has been added to the copy in order to record the results.

Figure 6.21 shows another example of another execution of the Test Case that was executed successfully in Figure 6.20. In this case, however, the Test Case execution fails at the point where the Safety Officer issues the 'check' *message*. Of course, any extra information concerning the failure may be included the in *note*.

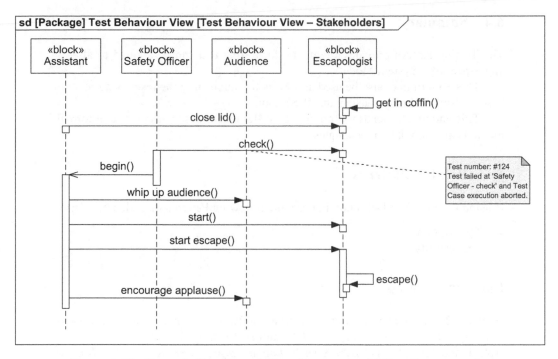

*Figure 6.20    Example Test Record View for a successful Test Case execution*

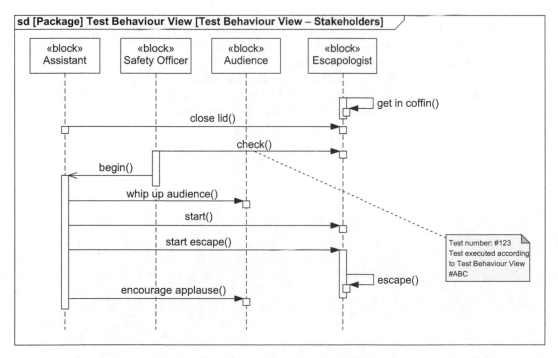

*Figure 6.21    Example Test Record View for unsuccessful Test Case execution*

## 6.4   Summary

The Testing Pattern provides a template for testing that can be applied to almost any aspect of a System Model.

This Pattern may also be used as a basis for automating the tests using model-based testing techniques using an MBSE tool.

This Pattern also lends itself to non-SysML modelling such as the use of formal methods and simulation techniques.

## 6.5   Related Patterns

If using the Interface Definition Pattern, the following Patterns may also be of use:

- Description
- Traceability.

## Further readings

Boehm, B. W. Verifying and Validating Software Requirements and Design Specifications. *IEEE Software1.1*. Jan/Feb 1984, pp. 75–88.

Holt, J. and Perry, S. *SysML for Systems Engineering; 2nd Edition: A Model-based Approach*. London: IET Publishing; 2013.

Object Management Group. *SysML website* [online]. Available at: http://www. omgsysml.org [Accessed February 2015].

Holt, J. *UML for Systems Engineering – Watching the Wheels*. 2nd Edition. London: IEE Publishing; 2004 (reprinted IET Publishing; 2007).

Rumbaugh, J., Jacobson, I. and Booch, G. *The Unified Modeling Language Reference Manual*. 2nd Edition. Boston, MA: Addison-Wesley; 2005.

Stevens, R., Brook, P., Jackson, K. and Arnold, S. *Systems Engineering – Coping with Complexity*. London: Prentice Hall Europe; 1998.

*Chapter 7*

# Epoch Pattern

## 7.1 Introduction

Measurement is a key aspect of any modelling endeavour. In order to manage and control the model it is essential that we can reason about it and, in order to reason about the model we need to be able to measure in a qualitative way, quantitative way or, more practically, both.

The Epoch Pattern allows measurements and metrics to be defined and applied to the model at specific points during the evolution of the model's life. These specific points in time are defined as Epochs and the reasoning is enabled by Measures and Metrics.

### 7.1.1 Pattern aims

The main aims of this Pattern are shown in the Architectural Framework Context View (AFCV) in Figure 7.1.

Figure 7.1 shows the context for the Epoch Pattern. The main aim of the Epoch Pattern is to understand the evolution of a system ('Understand evolution') by considering the evolution of the complexity of the system over time ('Support evolution of system through time').

There is a single high-level constraint that makes an assumption that a model of the System must exist ('Ensure model exists'). All of the Measure and Metrics that are defined using the Epoch Pattern Views are applied to the System model, therefore if it does not exist then no Measure nor Metrics may be applied.

There are three main use cases that must be satisfied for the Epoch Pattern to be effective which are:

- 'Define epoch', where a point in time is identified and an Epoch is defined at that point in time.
- 'Identify key viewpoints', where the System Model is looked at and a number of Viewpoints are identified that have been deemed relevant for the Epoch ('Identify relevant viewpoints'). Alongside the Viewpoints, it is also important that the Context of these Viewpoints is also known ('Identify contexts').
- 'Measure model', which includes defining Measure and Metrics ('Define general Metrics') and also applying them to the System Model ('Apply Metrics'). The results of the application of the Metrics must also be presented

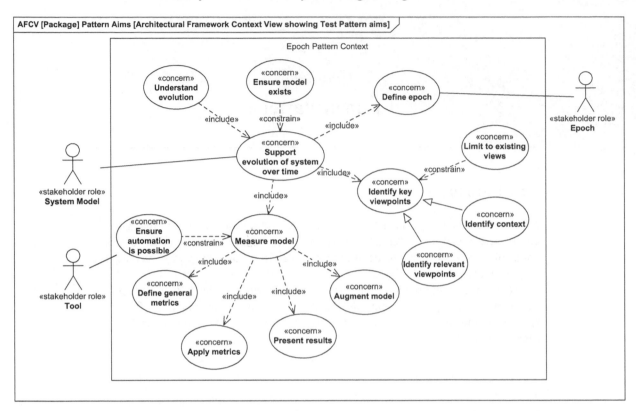

*Figure 7.1   Architectural Framework Context View showing Epoch Pattern aims*

to the modeller ('Present results') and then the model must be allowed to be updated or augmented based on these results ('Augment model'). There is a constraint that is applied to the 'Measure model' use case which is that automation of the application of the Metrics must be possible ('Ensure automation is possible').

This context drives the Epoch Pattern.

## 7.2   Concepts

The main concepts covered by the Epoch Pattern are shown in the Ontology Definition View (ODV) in Figure 7.2.

The Ontology for the Epoch pattern is shown in Figure 7.2 and describes the concepts and terminology, as follows:

- 'Epoch' that defines a point during the Life Cycle of the Project or System that will form a reference point for Project activities. An Epoch will apply to a specific baseline of the System Model, but not all baselines will be an Epoch. Epochs may be evolved from other Epochs, and may evolve into other Epochs.

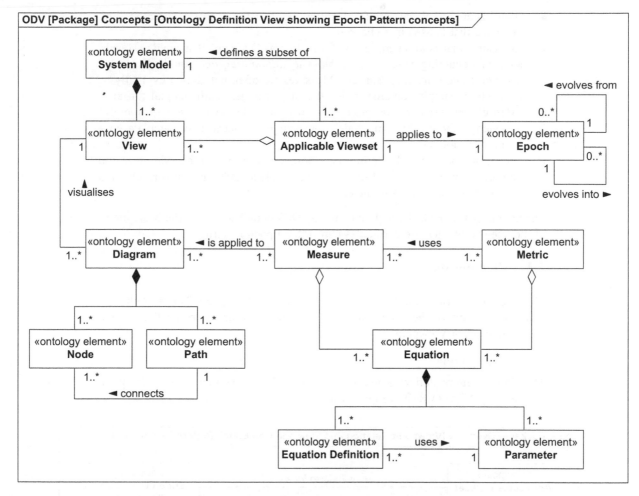

*Figure 7.2   Ontology Definition View showing Epoch concepts*

- 'Applicable Viewset' that is a subset of the 'System Model' and comprises one or more 'View'. These Views that make up the Epoch will be visualised by one or more 'Diagram'.
- 'System Model' is the abstraction of a specific System that evolves during the Life Cycle of the Project or System.
- 'View' is a defined set of information (based on a Viewpoint) with a standard structure that may be realised with one or more 'Diagram'. The Views make up the 'System Model'.
- 'Diagram' that is a visualisation of a 'View' and that uses a particular notation, in the case of this book, the language will be SysML.
- 'Node' is any graphical node (typically a shape) that forms part of the modelling language that is used in a Diagram.

- 'Path' is any graphical path (typically a line) that forms part of the modelling language that is used in a Diagram.
- 'Measure' that is an operation performed on a 'Diagram' that yields a result, such as a counting of *blocks* on a *block definition diagram*, counting *interactions* on a *sequence diagram*, etc. Measures are often not shown explicitly as they rely on simple counting, rather than any complex mathematical formula.
- 'Metric' that uses one or more 'Measure' in order to provide meaningful knowledge about a System Model, for example a Metric may calculate coupling, cohesion, etc.
- 'Equation' that forms the definition of both the 'Metric' and the 'Measure' and comprises an 'Equation Definition' (for example a mathematical formula) and its set of one or more 'Parameter'.

The Ontology Elements defined on Ontology Definition View form the basis for all of the Viewpoints that are defined as part of the Epoch Pattern.

## 7.3 Viewpoints

This section describes the Viewpoints that make up the Epoch Pattern. It begins with an overview of the Viewpoints, defines Rules that apply to the Pattern and then defines each Viewpoint.

### 7.3.1 Overview

The Epoch Pattern defines a number of Viewpoints as shown in the Viewpoint Relationship View (VRV) in Figure 7.3.

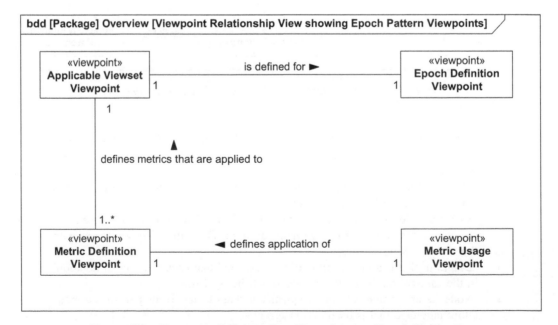

*Figure 7.3 Viewpoint Relationships View showing Epoch Viewpoints*

The Epoch Pattern defines four Viewpoints that are described as follows:

- The 'Epoch Definition Viewpoint' that identifies and describes the Epoch that is being defined.
- The 'Applicable Viewset Viewpoint' that identifies the subset of the System Model that is relevant to the Epoch being defined.
- The 'Metric Definition Viewpoint' where all of the Metrics and Measures are defined.
- The 'Metric Usage Viewpoint' that shows how the Metrics and Measures (defined in the Metric Definition Views) are applied to the Applicable Viewset (defined in the Applicable Viewset Views).

Each of these Viewpoints is described in more detail in the following sections. For each Viewpoint an example is also given.

### 7.3.2 Rules

Six rules apply to the four Epoch Viewpoints, as shown in the Rules Definition View (RDV) in Figure 7.4.

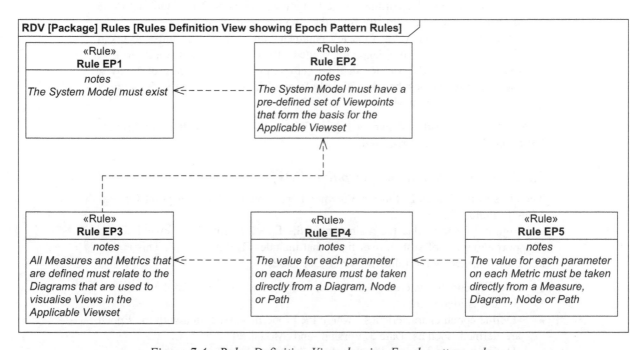

*Figure 7.4 Rules Definition View showing Epoch pattern rules*

Note that the five Rules shown in Figure 7.4 are the minimum that are needed, and are described at a high level as follows:

- 'Rule EP1' – 'The System Model must exist.' This Rules enforces the fundamental assumption that a System Model must exist as all of the Measure and

Metrics that are used as part of the Epoch Pattern are applied to the System Model. Therefore, if there is no System Model, then the Measure and Metrics cannot be applied.

- 'Rule EP2' – 'The System Model must have a pre-defined set of Viewpoints that form the basis for the Applicable Viewset.' This Rule follows on from EP1 as, not only must a model exist, but it must be structured into a set of Viewpoints that describe its Views. It is these Viewpoints and Views that are used to identify the Applicable Viewset.
- 'Rule EP3' – 'All Measures and Metrics that are defined must relate to the Diagrams that are used to visualise Views in the Applicable Viewset.' This Rule follows on from EP1 and EP2 and states that the Measures and Metrics that are defined must be applied to Diagrams that exist in the Applicable Viewset.
- 'Rule EP4' – 'The value for each parameter on each Measure must be taken directly from a Diagram, Node or Path.' This Rule follows on from EP3 and enforces the fact that each parameter for a Measure must be taken directly from a Diagram, Node or Path. When combined with Rule EP3, it ensures that the Measures are valid according to the Diagrams, Nodes or Paths, and that these themselves are taken directly from the Applicable Viewset.
- 'Rule EP5' – 'The value for each parameter on each Metric must be taken directly from a Measure, Diagram, Node or Path.' This Rule is similar to EP4 except here it is the Metrics that are being enforced. In the case of Metrics, their parameters must be taken directly from a Diagram, Node or Path, but may also be taken from a Measure.

Note that the five rules shown in Figure 7.4 are the minimum that are needed. Others could be added if required.

### 7.3.3   Epoch Definition Viewpoint (EDVp)

The aims of the Epoch Definition Viewpoint are shown in the Viewpoint Context View in Figure 7.5.

Figure 7.5 shows that the main aim of the Epoch Definition Viewpoint is to 'Support evolution of system over time' that includes 'Define epoch'. This is broken down further by the following three Use Cases:

- 'Identify epoch', where a specific point during the Life Cycle of the Project or System is identified.
- 'Define epoch characteristics', where the properties associated with the Epoch are defined, such as 'Date', 'Version' and 'Model ref'.
- 'Define relationships to other epochs', which entails looking at other epochs in the Life Cycle and identifying relationship between them and the new epoch. This will allow the evolution of the model to be explored.

It should be stressed that the aim of this Viewpoint is to define the Epochs that show the evolution of the System Model over time, but not to define exactly what aspect of the System Model is being measured (such as size, complexity).

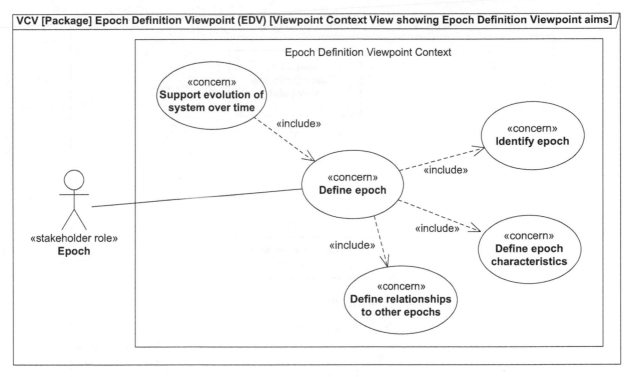

*Figure 7.5   Viewpoint Context View showing Epoch Definition Viewpoint aims*

The elements that are relevant for the Epoch Definition Viewpoint are shown in the next section.

### 7.3.3.1   Description

The Viewpoint Definition View (VDV) in Figure 7.6 shows the Ontology Elements that appear on an Epoch Definition Viewpoint.

Figure 7.6 shows the elements from the Ontology that are relevant to the Epoch Definition Viewpoint. The Epoch Definition Viewpoint shows the Epoch being defined with its property values filled in and the relationships between them.

An Epoch may be related to other Epochs in two ways:

• By evolving from another Epoch, where there is an existing Epoch.
• By evolving into another Epoch, where there will be a future Epoch.

Each Epoch is described by three *properties*, which are:

• 'Baseline' the version number for the Baseline that the Epoch represents.
• 'Date' the date at which the Epoch was created.
• 'Model Reference' that refers to a specific artefact in the System model.

The number of *properties* defined for each Epoch is not limited to the three listed here, but these should be taken as a minimum set.

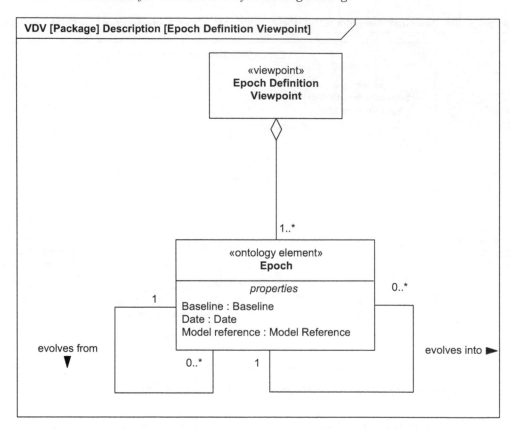

*Figure 7.6    Viewpoint Definition View showing the elements that appear on the Epoch Definition Viewpoint (EDVp)*

### 7.3.3.2    Example

An example View that conforms to the Epoch Definition Viewpoint is shown in Figure 7.7.

The Epoch Definition View in Figure 7.7, realised as a *block definition diagram*, shows two *instance specifications* of Epoch, 'Epoch 1' and 'Epoch 2' that are connected together by the *link*.

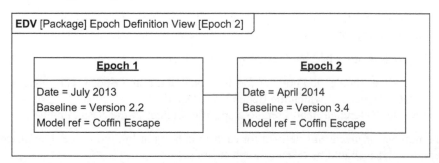

*Figure 7.7    An example of an Epoch Definition View showing two Epochs*

In the example shown here, two Epochs have been identified: 'Epoch 1' and 'Epoch 2'.

Any number of Epochs may be shown using this Viewpoint, ranging from a single Epoch to a string of Epochs that are sequenced together. It should also be noted that the Epochs need not be connected in a linear fashion, it is also possible to have branches to show divergence in the evolution of the System Model.

### 7.3.4   Applicable Viewset Viewpoint (AVVp)

The aims of the Applicable Viewset Viewpoint are shown in the Viewpoint Context View in Figure 7.8.

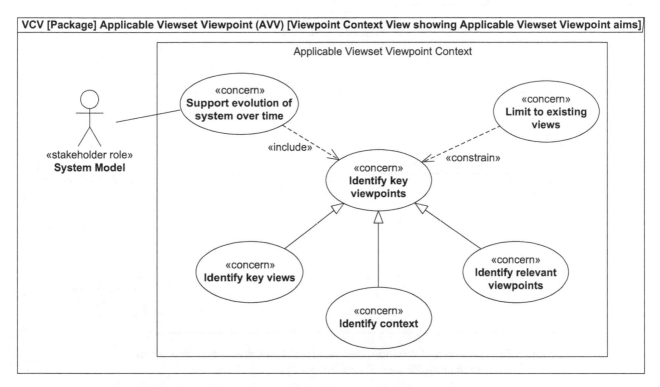

*Figure 7.8   Viewpoint Context View showing Applicable Viewset Viewpoint aims*

Figure 7.8 shows that the main aim of the Applicable Viewset Viewpoint is to 'Support the evolution of system over time' which includes 'Identify key viewpoints', which has three types associated with it:

- 'Identify relevant viewpoint', that requires the subset of the System Model that is relevant for the Epoch being defined to be identified.
- 'Identify context', where the Context refers to the Context of the Viewpoints that have been identified (hence the constraint between the two *use cases*). This is important as it is possible that the Context for those Viewpoints has evolved as time has gone on and, hence, this evolution must be captured.

● 'Identify key views', where these Views will be the specific instances of the Viewpoints that have been identified.

There is also a constraint, 'Limit to existing views' that really enforces Rules 'Rule EP1' and 'Rule EP2' from Figure 7.4.

### 7.3.4.1   Description

The VDV in Figure 7.9 shows the Ontology Elements that appear on an Applicable Viewset Viewpoint.

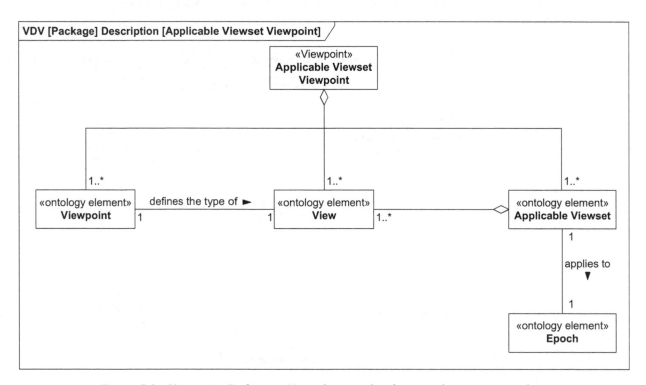

*Figure 7.9    Viewpoint Definition View showing the elements that appear on the Applicable Viewset Viewpoint (AVVp)*

The Applicable Viewset Viewpoint shows how the 'Applicable Viewset' is related to a specific 'Epoch' and comprises a one or more 'View' and one or more 'Viewpoint'.

In the case that the Epoch refers to the whole model, then this diagram becomes very simple as it need only show a single reference to the model, however, in many cases the Epoch will refer to a subset of the model. In this situation, it is important to which Viewpoint and specific associated Views are relevant.

### 7.3.4.2   Example

An example View that conforms to the Applicable Viewset Viewpoint is shown in Figure 7.10.

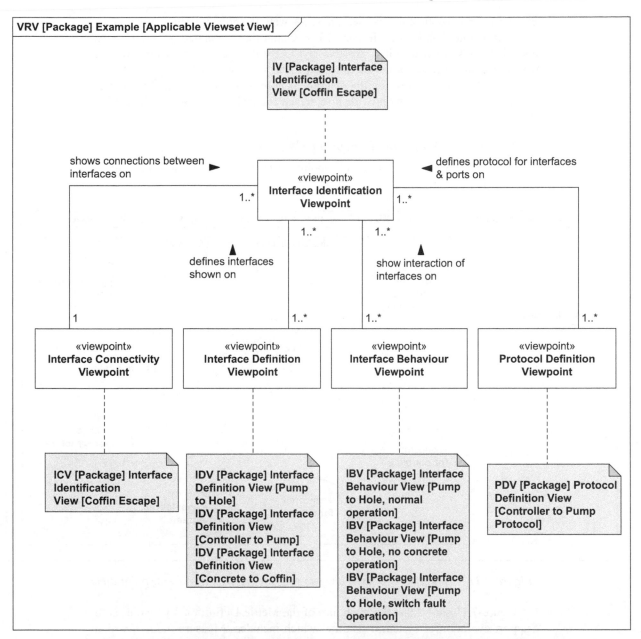

*Figure 7.10    An example of an Applicable Viewset View showing Service-Based Interfaces*

The Applicable Viewset View in Figure 7.10, realised as a SysML *block definition diagram*, shows the Views that are relevant to a specific Epoch. In this example, the Viewpoints are the ones that are defined by the Interface Pattern. Note how the Interface Pattern is being re-used here to define which Viewpoints are relevant to this Applicable Viewset View. As well as knowing which Viewpoints form part of the Applicable Viewset, it is essential that the actual Views are also identified. This is because it is possible and likely that neither the entire System Model, nor an entire set of instances of a specific Viewpoint will be relevant.

These Views are shown on the diagram as *notes* that are associated with their relevant Viewpoint and that provide a reference for each View that is relevant for each Viewpoint.

### 7.3.5   Metric Definition Viewpoint (MDVp)

The aims of the Metric Definition Viewpoint are shown in the Viewpoint Context View in Figure 7.11.

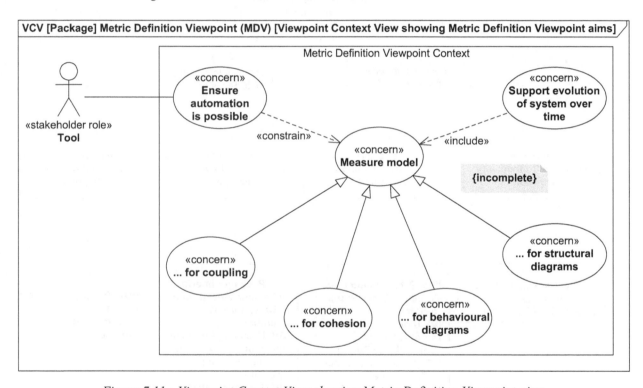

*Figure 7.11   Viewpoint Context View showing Metric Definition Viewpoint aims*

Figure 7.11 shows that the main aim of the Metric Definition Viewpoint is to 'Support evolution of system over time' which includes 'Measure model', which includes 'Define general Metrics' that has four types:

- '... for coupling' that is a generalisation relating to complexity. This requires that general Metrics are defined that measure and calculate some aspect of the complexity between Nodes.
- '... for cohesion' that is a *generalisation* relating to complexity. This requires that general Metrics are defined that measure and calculate some aspect of the complexity inside a Node.

- '... for structural diagrams' that is a *generalisation* related to an aspect of the System Model. This requires that the general Metrics can be applied to the structural aspect of the System Model.
- '... for behavioural diagrams' that is a *generalisation* related to an aspect of the System Model. This requires that the general Metrics can be applied to the behavioural aspect of the System Model.

It should be noted that the four types of measurement that are included in this diagram provide examples of the type of measurements that are possible and the diagram is not intended to show an exhaustive list. This is represented by the '{incomplete}' constraint.

There is a general constraint that requires that any Metrics developed must be able to be automated ('Ensure automation is possible').

### 7.3.5.1   Description

The VDV in Figure 7.12 shows the Ontology Elements that appear on a Metric Definition Viewpoint.

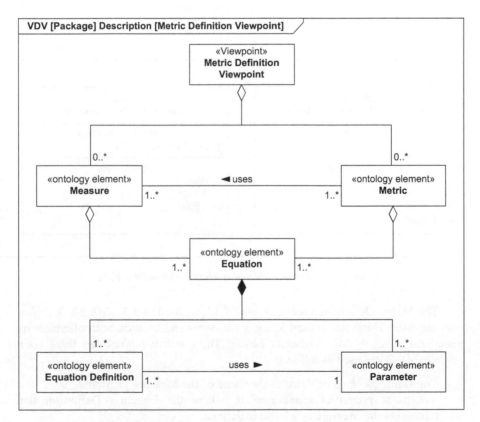

*Figure 7.12   Viewpoint Definition View showing the elements that appear on the Metric Definition Viewpoint (MDVp)*

Figure 7.12 shows the Metric Definition Viewpoint defines the Measures and the Metrics associated with them. Each Measure and Metric is defined here in terms of its Equations and associated Equation Definitions and Parameters.

Both the Measures and Metrics are described in the same way, buy using Equation. Each Equation is made up of a number of Equation Definitions that use a number of Parameters.

### 7.3.5.2   Example

An example View that conforms to the Metric Definition Viewpoint is shown in Figure 7.13.

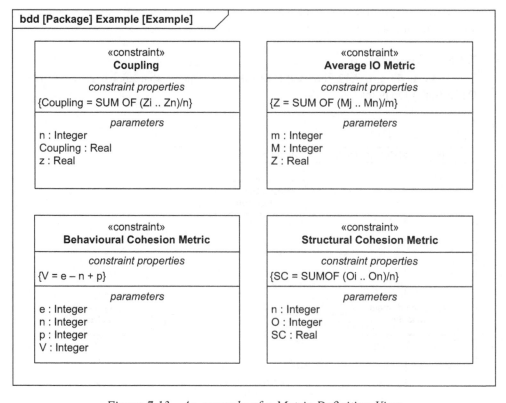

*Figure 7.13   An example of a Metric Definition View*

The Metric Definition View in Figure 7.13, realised as a SysML *block definition diagram*. This View is used to show Measures and Metrics, both of which are visualised using SysML *constraint blocks*. The *constraint block* has three *compartments* that are used as follows:

– Constraint *block* name – this is the name of the Measure or Metric.
– 'constraint properties' *compartment*, where the Equation Definition that represents the Measure or Metric is defined.

– 'parameters' *compartment*, where the Parameters that are used by the Equation Definition are defined according to their type and multiplicity.

Figure 7.13 shows four Metrics that are relevant to the Epoch used in this example. These Metrics are defined as follows.

– 'Coupling', that calculates the coupling between different nodes on a single diagram, and that uses the 'Average IO Metric' (Cruickshank and Gaffney,1980). The equation for this Metric is:

{Coupling = SUMOF (Zi ... Zn)/n}, where:

> Z is the average number of inputs and outputs for the node i
> i is the count of the nodes on the diagram
> n is the total number of nodes on the diagram.

– 'Average IO Metric', that calculates the average number of inputs and outputs shared over m nodes. The equation for the Metric is:

{Z= SUMOF (Mj ... Mn)/m}, where:

> M is the sum of the inputs and outputs shared by nodes j and i
> m is the total number of nodes to which i is joined.

– 'Structural Cohesion Metric', that calculates the average number of operations per block in a diagram (Lorenz and Kidd, 1994). The equation for the Metric is:

{SC = SUMOF (Oi ... On)/n}, where:

> O is the number of operations in a single block
> n is the number of blocks.

– 'Behavioural Cohesion Metric', that calculates the complexity of a behavioural diagram by calculating the possible number of paths through its decision structure (McCabe, 1976). The equation for the Metric is:

{V = e – n + p}, where:

> e is the number of nodes
> n is the number of paths
> p is the number of start and end points.

The Metrics that are defined here are used purely as an example of how established techniques may be applied to the Applicable Viewset for a specific Epoch. These Metrics were traditionally applied at the code level in software engineering in the original references, but have been tailored to be able to be applied at the model level – see Tugwell (2001) for more information.

In the example shown here there are no explicit Measures being defined, as these Measures are simple counts on the diagram. For example, the 'Behavioural Cohesion Metric' defined here is using some simple counts that look at a diagram and identify how many nodes exist. These simple counts are the Measures.

### 7.3.6   Metric Usage Viewpoint (MUVp)

The aims of the Metric Usage Viewpoint are shown in the Viewpoint Context View in Figure 7.14.

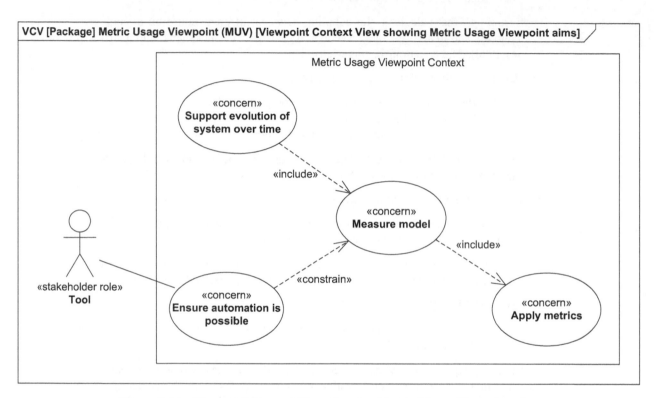

*Figure 7.14   Viewpoint Context View showing Metric Usage Viewpoint aims*

Figure 7.14 shows that the main aim of the Metric Usage Viewpoint is to 'Support evolution of system over time' which includes 'Measure model', which includes being able to apply the Metrics to the System Model ('Apply Metrics').

There is a general constraint that requires that any Metrics developed must be able to be automated ('Ensure automation is possible').

#### 7.3.6.1   Description

The VDV in Figure 7.15 shows the Ontology Elements that appear on a Metric Usage Viewpoint.

The Metric Usage Viewpoint in Figure 7.15 shows how the Measures and Metrics are applied to the System Model via the Diagrams.

The Diagrams that the Measures and, hence, Metrics are applied to must have already been identified as part of the Applicable Viewset View.

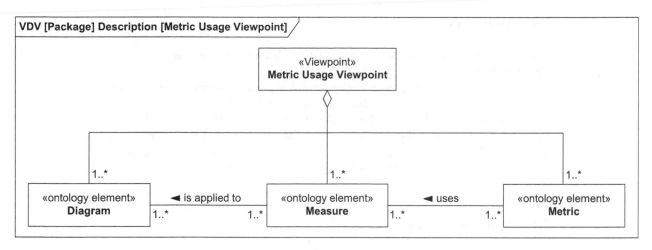

*Figure 7.15    Viewpoint Definition View showing the elements that appear
on the Metric Usage Viewpoint (MUVp)*

### 7.3.6.2    Example

Example Views that conform to the Metric Usage Viewpoint are shown in
Table 7.1.

The Metric Usage View is realised as simple table as shown in Table 7.1. The
table relates the Metrics and Measures that were defined in the Metrics Definition
View to the relevant Views in the Applicable Viewset View.

*Table 7.1    An example of the Metric Usage View*

| Measure/Metric | Applied to |
|---|---|
| Coupling (using Average IO Metric) | IIV [Package] Interface Identification View [Coffin Escape]<br>ICV [Package] Interface Identification View [Coffin Escape]<br>IDV [Package] Interface Definition View [Pump to Hole]<br>IDV [Package] Interface Definition View [Controller to Pump]<br>IDV [Package] Interface Definition View [Concrete to Coffin] |
| Structural Cohesion Metric | IIV [Package] Interface Identification View [Coffin Escape]<br>ICV [Package] Interface Identification View [Coffin Escape]<br>IDV [Package] Interface Definition View [Pump to Hole]<br>IDV [Package] Interface Definition View [Controller to Pump]<br>IDV [Package] Interface Definition View [Concrete to Coffin] |
| Behaviour Cohesion Metric | IBV [Package] Interface Behaviour View [Pump to Hole, normal operation]<br>IBV [Package] Interface Behaviour View [Pump to Hole, no concrete operation]<br>IBV [Package] Interface Behaviour View [Pump to Hole, switch fault operation]<br>PDV [Package] Protocol Definition View [Controller to Pump Protocol] |

## 7.4   Summary

This section has introduced and defined the set of Viewpoints that that make up the Epoch Pattern.

The Epoch Pattern allows modellers to define a set of Measures and Metrics that may then be applied to a Model, or sub-set of a Model.

It should be noted that the Epoch Pattern provides a mechanism that allows modellers to define Measures and Metrics, but the pattern itself does not define any Metrics or Measures. The pattern itself is purely definitional in nature. An example set of Metrics is shown here based on previous work on applying Metrics and Measures to Models.

## 7.5   Related Patterns

When using the Epoch Pattern, the following patterns may also be of use:

- Description Pattern
- Traceability Pattern
- Life Cycle Pattern.

## References

Cruickshank, R.D. and Gaffney, J.E., Jr., *Software Design Coupling and Strength Metrics*. Greenbelt, Maryland: NASA, Goddard Space Flight Centre; 1980.

Lorenz, M. and Kidd, J. *Object-Oriented Software Metrics*. Prentice Hall, New Jersey, USA, 1994.

McCabe T. 'A Complexity Measure", *IEEE Transactions on Software Engineering*, Vol. SE2, No. 4, pp. 308–320, 1976.

Tugwell, G.C. 'Metrics for Full Systems Engineering Lifecycle Activities (MeFuSELA)'. PhD Thesis, University of Wales Swansea, 2001.

## Further reading

Holt, J. and Perry, S. *SysML for Systems Engineering; 2nd Edition: A Model-Based Approach*. London: IET Publishing; 2013.

*Chapter 8*

# Life Cycle Pattern

## 8.1 Introduction

All systems that are developed pass through a number of Life Cycles. The nature of the various Life Cycles that a system goes through are often misunderstood. There are many different types of Life Cycle that exist, including, but not limited to:

- *Project Life Cycles*. Project Life Cycles are perhaps, along with Product Life Cycles, one of the most obvious examples of applications of Life Cycles. We tend to have rigid definitions of the terminal conditions of a project, such as start and end dates, time scales, budgets and resources and so a Project Life Cycle is one that many people will be able to identify with.
- *Product Life Cycles*. Another commonly considered Life Cycle is that of the product (system) itself. It is relatively simple to visualise the conception, development, production, use, support and disposal of a product.
- *Programme Life Cycles*. Most projects will exist in some sort of higher-level programme. Each of the programmes will also have its own Life Cycle which will have some constraints on the Life Cycles of all projects that are contained within it.
- *System Procurement Life Cycles*. Some systems may have a procurement Life Cycle that applies to them. From a business point of view, this may be a better way to view a product, or set of products, than looking at the Product Life Cycle alone.
- *Technology Life Cycles*. Any technology will have a Life Cycle. For example, in the past, the accepted norm for removable storage was magnetic tapes. This was then succeeded by magnetic discs, then optical discs, then solid-state devices and then virtual storage. Each of these technologies has its own Life Cycle.
- *Equipment Life Cycles*. Each and every piece of equipment will have its own Life Cycle. This may start before the equipment is actually acquired and may end when the equipment has been safely retired. Stages of the equipment Life Cycle may describe its current condition, such as whether the equipment is in use, working well, degrading and so on.
- *Business Life Cycles*. The business that acquires, engineers, project manages or uses a system will have a Life Cycle. In some cases, the main driver of the

business may be to remain in business for several years and then to sell it on. Stages of the Life Cycle may include expansion or growth, steady states, controlled degradation and so on.

These different types of Life Cycle not only exist, but often interact in complex ways. Also, each of these Life Cycles will control the way that processes are executed during each stage in the Life Cycle. The Life Cycle Pattern is intended to allow the definition, behaviour and interaction of the various Life Cycles of a system to be defined and understood

### 8.1.1   Pattern aims

This pattern is intended to be used as an aid to the understanding of the Life Cycles involved in the MBSE development of a system. The main aims of this pattern are shown in the Architectural Framework Context View (AFCV) in Figure 8.1.

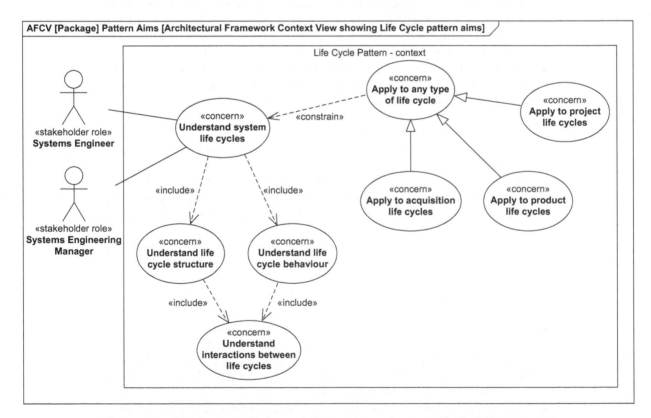

*Figure 8.1   Architectural Framework Context View showing Life Cycle Pattern aims*

The key aim of the Life Cycle Pattern is to allow a 'Systems Engineer' and a 'Systems Engineering Manager' to 'Understand system life cycles'. This key aim is constrained by the need to 'Apply to any type of life cycle', including (but not limited to):

- Project Life Cycles ('Apply to project life cycles')
- Product Life Cycles ('Apply to product life cycles')
- Procurement Life Cycles ('Apply to procurement life cycles').

The main need to 'Understand system life cycles' includes the need to understand a life cycle in term of its structure ('Understand lifecycle structure') and its behaviour ('Understand life cycle behaviour'). Both of these include the need to 'Understand interactions between life cycles', allowing multiple Life Cycles for a System to be related one to another.

## 8.2 Concepts

The main concepts covered by the Life Cycle Pattern are shown in the Ontology Definition View (ODV) in Figure 8.2.

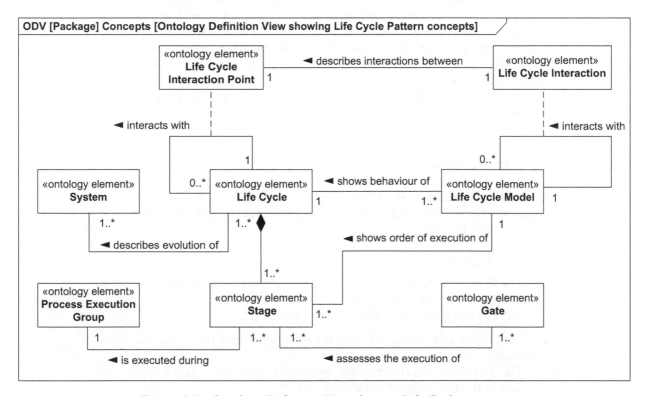

*Figure 8.2   Ontology Definition View showing Life Cycle concepts*

The main concept is that of a 'Life Cycle', a set of one or more 'Stage' that can be used to describe the evolution of System over time Each 'Stage' represents a period within a 'Life Cycle' that relates to its realisation through one or more 'Process Execution Group' (see discussion below). The success of a 'Stage' is assessed by one or more 'Gate', a mechanism for assessing the success or failure of the execution of a 'Stage'. These Ontology Elements allow the structural aspects of a Life Cycle to be captured.

The behavioural aspects are represented by the concept of a 'Life Cycle Model' which represents the ordered execution of a set of one or more 'Stage', showing the behaviour of a 'Life Cycle'.

It is important to understand the difference between a Life Cycle and a Life Cycle Model:

- A Life Cycle is a structural construct that shows the Stages that make up a Life Cycle.
- A Life Cycle Model is a behavioural construct that shows how the Stages behave within the execution of a Life Cycle.

A 'Life Cycle' may interface with others. This is done through a 'Life Cycle Interaction Point'. Each corresponding 'Life Cycle Model' will then interact with others through a 'Life Cycle Interaction', the point during a 'Life Cycle Model' at which they interact, reflected in the way that one or more 'Stage' interact with each other.

The concept of the 'Process Execution Group, an ordered execution of one or more Process that is performed as part of a 'Stage', is a linking concept that can be used to relate the above Life Cycle concepts to those of process modelling and project planning. This is discussed further in the section on the Life Cycle Model Viewpoint below. For a full discussion see Holt and Perry (2013).

## 8.3    Viewpoints

This section describes the Viewpoints that make up the Life Cycle Pattern. It begins with an overview of the Viewpoints, defines Rules that apply to the pattern and then defines each Viewpoint.

### 8.3.1    Overview

The Life Cycle Pattern defines a number of Viewpoints as shown in the Viewpoint Relationship View (VRV) in Figure 8.3.

The Life Cycle pattern defines four Viewpoints:

- The 'Life Cycle Viewpoint' that identifies the Stages that exists in a Life Cycle.
- The 'Life Cycle Model Viewpoint' that describes how each Stage behaves over time in relation to one or more other Stage.
- The 'Interaction Identification Viewpoint' that identifies the Life Cycle Interaction Points between one or more Life Cycle.

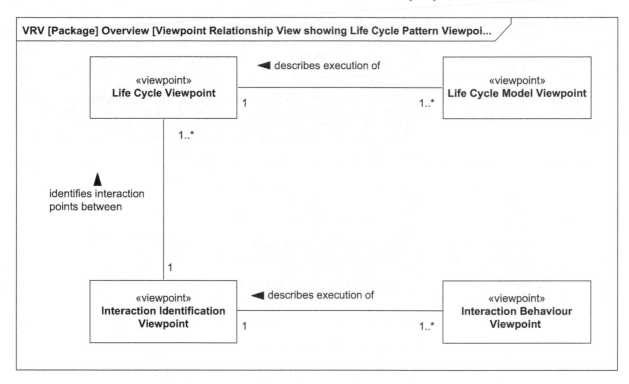

*Figure 8.3   Viewpoint Relationship View showing Life Cycle Viewpoints*

- The 'Interaction Behaviour Viewpoint' that describes the execution of each Life Cycle Interaction Point in relation to one or more other Life Cycle Interaction Point as identified in an Interaction Identification Viewpoint.

Each of these Viewpoints is described in more detail in the following sections. For each Viewpoint an example View that conforms to the Viewpoint is also given.

### 8.3.2   Rules

Five Rules apply to the four Pattern Viewpoints, as shown in the Rules Definition View (RDV) in Figure 8.4.
   The Life Cycle Pattern has five Rules defined, which are:

- 'LC1: Each instance of the Life Cycle Viewpoint must consist of at least one Stage.' This is a basic common-sense check that ensures that no Life Cycles are defined that are empty in that they have no Stages.
- 'LC2: Each instance of the Life Cycle Model Viewpoint must be based on a corresponding instance of the Life Cycle Viewpoint that shows the correct Stages.' This Rules represents a consistency check that ensures that all Life Cycle Model Views (behavioural) are based on an existing Life Cycle (structural).

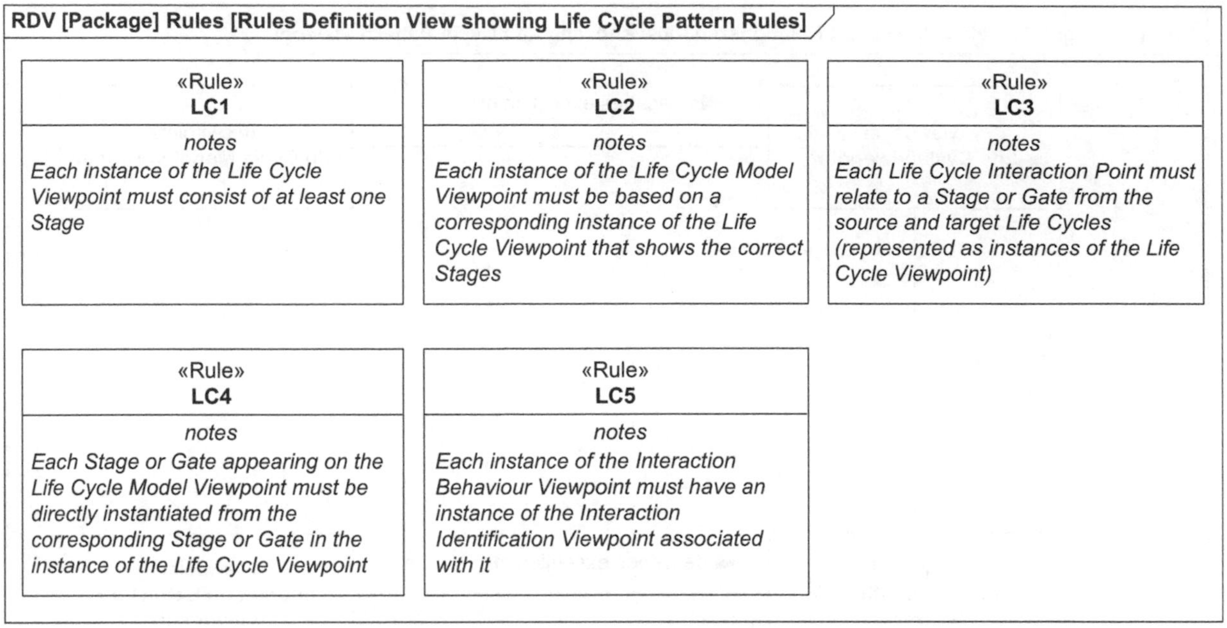

RDV [Package] Rules [Rules Definition View showing Life Cycle Pattern Rules]

| «Rule» **LC1** | «Rule» **LC2** | «Rule» **LC3** |
|---|---|---|
| *notes* | *notes* | *notes* |
| *Each instance of the Life Cycle Viewpoint must consist of at least one Stage* | *Each instance of the Life Cycle Model Viewpoint must be based on a corresponding instance of the Life Cycle Viewpoint that shows the correct Stages* | *Each Life Cycle Interaction Point must relate to a Stage or Gate from the source and target Life Cycles (represented as instances of the Life Cycle Viewpoint)* |

| «Rule» **LC4** | «Rule» **LC5** |
|---|---|
| *notes* | *notes* |
| *Each Stage or Gate appearing on the Life Cycle Model Viewpoint must be directly instantiated from the corresponding Stage or Gate in the instance of the Life Cycle Viewpoint* | *Each instance of the Interaction Behaviour Viewpoint must have an instance of the Interaction Identification Viewpoint associated with it* |

*Figure 8.4   Rules Definition View showing Life Cycle rules*

- 'LC3: Each Life Cycle Interaction Point must relate to a Stage or Gate from the source and target Life Cycles (represented as instances of the Life Cycle Viewpoint).' This Rules represents a consistency check that ensure that all Interaction Points have a source and target that relate to Stages within the various Life Cycles.
- 'LC4: Each Stage or Gate appearing on the Life Cycle Model Viewpoint must be directly instantiated from the corresponding Stage or Gate in the instance of the Life Cycle Viewpoint.' This Rule represents another consistency check that ensures that the elements on the Life Cycle Model Views (the Gates and Stage) are taken from existing definitions from the Life Cycle View.
- 'LC5: Each instance of the Interaction Behaviour Viewpoint must have an instance of the Interaction Identification Viewpoint associated with it.' This Rules represents another consistency check that ensures that the Interactions on the Interaction Behaviour Views have related Interaction Points on the Interaction Point Identification Views.

Note that the five Rules shown in Figure 8.4 are the minimum that are needed. Others could be added if required.

### 8.3.3   Life Cycle Viewpoint (LCVp)

The aims of the Life Cycle Viewpoint are shown in the Viewpoint Context View in Figure 8.5.

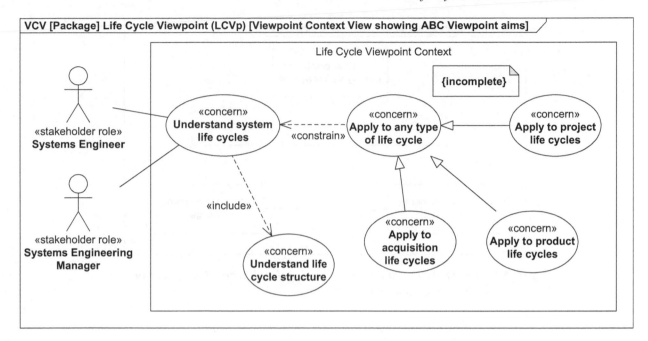

*Figure 8.5   Viewpoint Context View showing Life Cycle Viewpoint aims*

The main aim of the Life Cycle Viewpoint is to 'Understand system life cycles' in terms of their structure ('Understand life cycle structure'), in a way that can 'Apply to any type of life cycle'.

There are three types of Life Cycle shown here, which are:

- Project Life Cycles ('Apply to project life cycles'),
- Product Life Cycles ('Apply to product life cycles') and
- Acquisition Life Cycles (Apply to acquisition life cycles).

This is not intended to be an exhaustive list, as indicated by the '{incomplete}' constraint.

### 8.3.3.1   Description

The Viewpoint Definition View (VDV) in Figure 8.6 shows the Ontology Elements that appear on a Life Cycle Viewpoint.

The Life Cycle Viewpoint is a relatively simple Viewpoint, showing the structure of a single Life Cycle in terms of the Stages that it is made up of.

### 8.3.3.2   Example Life Cycle Viewpoint

An example View that conforms to the Life Cycle Viewpoint is shown in Figure 8.7.

The Life Cycle View in Figure 8.7, realised as a SysML *block definition diagram*, defines a Life Cycle for development based on the international standard

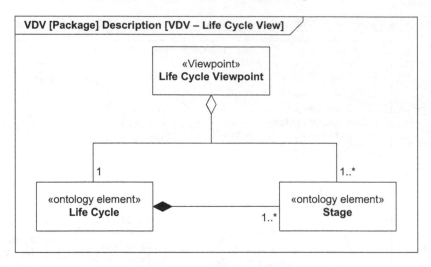

*Figure 8.6    Viewpoint Definition View showing the elements that appear on the
Life Cycle Viewpoint (LCVp)*

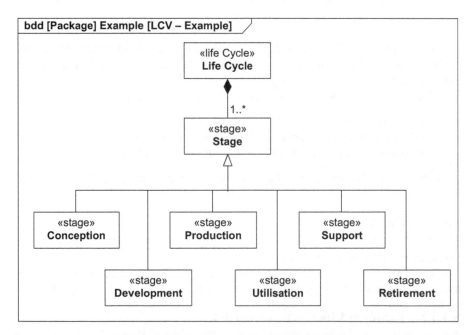

*Figure 8.7    An example Life Cycle View*

ISO15288:2008. It shows the 'Life Cycle' is made up of six 'Stage': 'Conception', 'Development', 'Production', 'Utilisation', 'Support' and 'Retirement'.

The Life Cycle and each of its Stages are represented by *blocks* on the *block definition diagram*.

This View fulfils Rule LC1.

### 8.3.4 *Life Cycle Model Viewpoint (LCMVp)*

The aims of the Life Cycle Model Viewpoint are shown in the Viewpoint Context View in Figure 8.8.

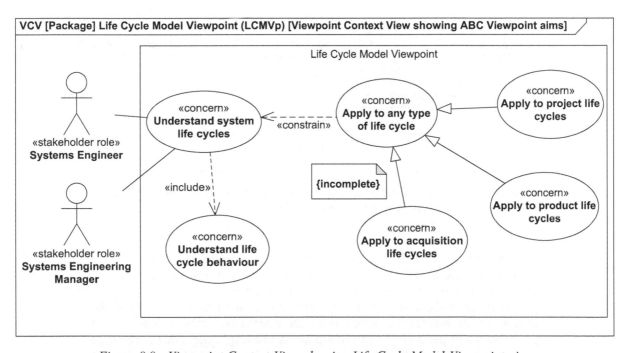

*Figure 8.8   Viewpoint Context View showing Life Cycle Model Viewpoint aims*

The main aim of the Life Cycle Model Viewpoint is to 'Understand system life cycle' that focuses on the behaviour – 'Understand life cycle behaviour'. This is constrained by being able to be applied to any types of Life Cycle ('Apply to any type of life cycle') and, again some examples of types of Life Cycle are shown, but there may be others, which is expressed by the '{incomplete}' constraint.

#### 8.3.4.1   Description

The VDV in Figure 8.9 shows the Ontology Elements that appear on a Life Cycle Model Viewpoint.

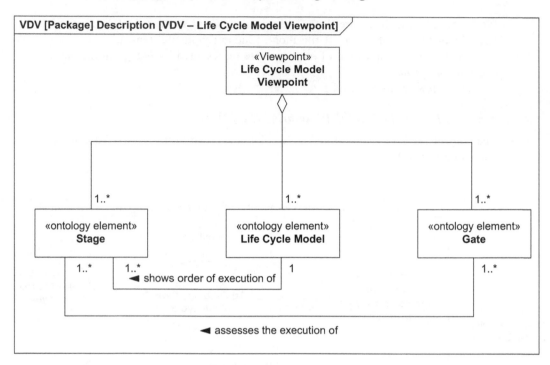

*Figure 8.9    Viewpoint Definition View showing the elements that appear on the*
*Life Cycle Model Viewpoint (LCMVp)*

The Life Cycle Model Viewpoint shows the Stages and Gates that make up a single Life Cycle Model, showing the behaviour of a Life Cycle through time and the Gates controlling behaviour at the end of each Stage.

As noted in the discussion on the Ontology for the Life Cycle Pattern, the concept of the Process Execution Group, an ordered execution of a number of processes that is performed as part of a Stage, is a linking concept that can be used to expand a Life Cycle Model to tie the concepts of process modelling to those of project planning. It allows the Life Cycle Pattern, through the Life Cycle Model Viewpoint, to bridge the gap often found in Systems Engineering between the defined Systems Engineering processes followed by the systems engineers developing a System and the project plans produced by project managers tasked with managing the development of the System. This is done by expanding the definition of the Life Cycle Model Viewpoint to allow Views based on the Viewpoint to show the Process Execution Groups that are executed in a particular Stage, along with the sequencing and timing of such executions. Further expansion allows the processes within a Process Execution Group to be shown in a similar fashion. Such changes would require additions to the Ontology and changes to the VDV for the Life Cycle Model Viewpoint. For a full discussion of this, see Holt and Perry (2013).

### 8.3.4.2   Example Life Cycle Model Viewpoint

An example View that conforms to the Life Cycle Model Viewpoint is shown in Figure 8.10.

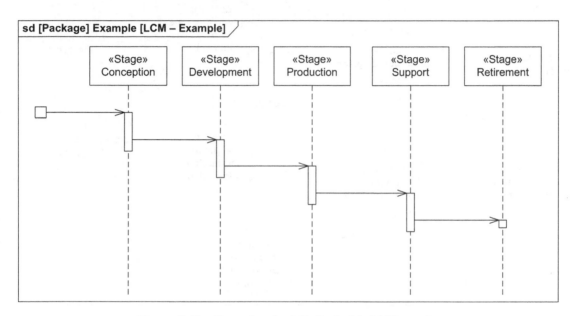

*Figure 8.10   Example of a Life Cycle Model Viewpoint*

The Life Cycle Model View in Figure 8.10, realised as a SysML *sequence diagram,* shows a typical development Life Cycle Model View based on the Life Cycle defined in Figure 8.7. Each Stage is represented by a *life line* and corresponds to one of the Stages represented using *blocks* on Figure 8.7. Each *life line* also shows an *execution specification* (the small rectangles) that may be annotated to include timing constraints or parallelism between various Stages.

If the expanded Life Cycle Model Viewpoint discussed above is being used, then additional Views would be produced, one for each Stage, showing the Process Execution Groups taking place within each Stage. A further set would typically also be produced, one for each Process Execution Group, showing the processes executed with each Process Execution Group. See Holt and Perry (2013) for a full discussion.

### 8.3.5   *Interaction Identification Viewpoint (IIVp)*

The aims of the Interaction Identification Viewpoint are shown in the Viewpoint Context View in Figure 8.11.

The main aim of the Interaction Identification Viewpoint is to 'Understand system life cycles' in terms of their interactions ('Understand interactions between

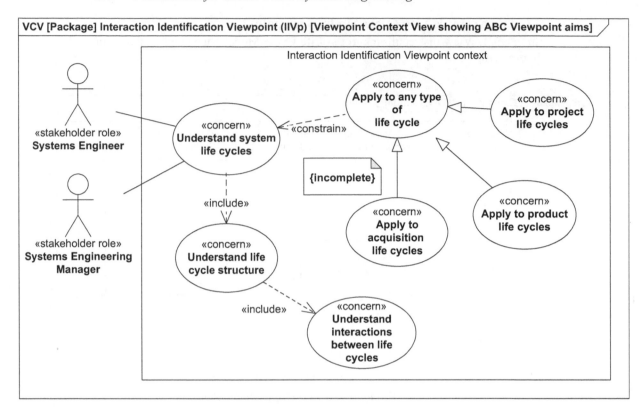

*Figure 8.11   Viewpoint Context View showing Interaction Identification Viewpoint aims*

life cycles') from a structural point of view ('Understand life cycle structure'), in a way that can 'Apply to any type of life cycle'.

The focus of this Viewpoint is on identifying the Interaction Points that exist between Life Cycles ('Understand interactions between life cycles').

### 8.3.5.1   Description

The VDV in Figure 8.12 shows the Ontology Elements that appear on a Interaction Identification Viewpoint.

The Interaction Identification Viewpoint shows one or more Life Cycles, the Stages that make them up and the Life Cycle Interaction Points between the Stages. Note that this shows structural information – it shows which Stages from which Life Cycles have interactions but not when they interact.

### 8.3.5.2   Example Interaction Identification Viewpoint

An example View that conforms to the Interaction Identification Viewpoint is shown in Figure 8.13.

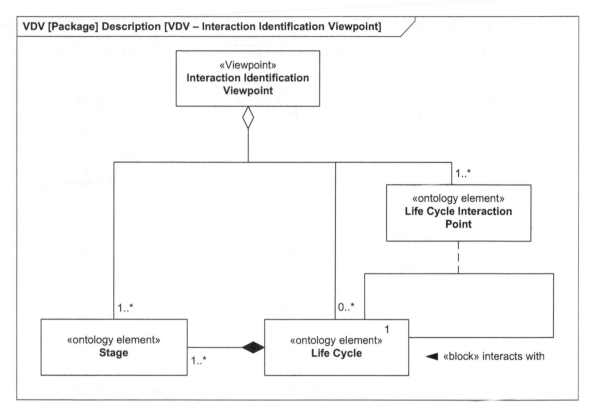

*Figure 8.12    Viewpoint Definition View showing the elements that appear on the
Interaction Identification Viewpoint (IIVp)*

The Interaction Identification View in Figure 8.13, realised as a SysML *block
definition diagram*, shows three Life Cycles, the 'Development Life Cycle',
'Acquisition Life Cycle' and 'Deployment Life Cycle'. Each of these Life Cycles is
represented using a *package*. The Stages in each Life Cycle are realised using
*blocks* with the stereotype «stage». Each Life Cycle Interaction Point is represented
as a *dependency* with the stereotype «interaction point».

For example, there is a Life Cycle Interaction Point between the 'Delivery' Stage
of the 'Acquisition Life Cycle' and the 'Conception' Stage of the 'Development Life
Cycle', represented by the «interaction point» *dependency* between them. Note that
the full term Life Cycle Interaction Point is not used here for the stereotype on the
*dependencies* purely for reasons of clarity and presentation.

The Interaction Identification View in Figure 8.13, along with the Life Cycle
View in Figure 7, fulfils Rule LC3.

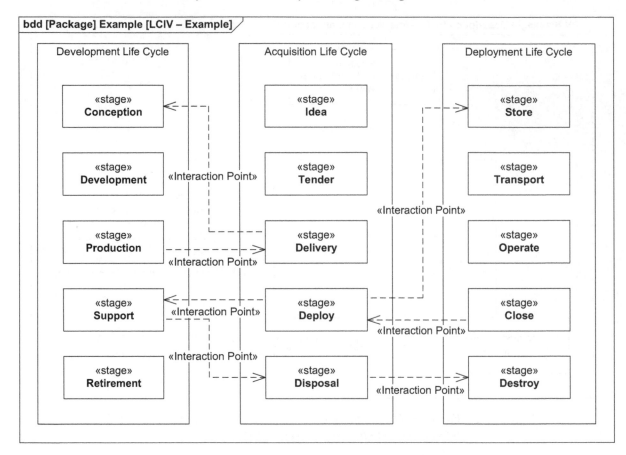

*Figure 8.13    An example of an Interaction Identification View*

### 8.3.6    *Interaction Behaviour Viewpoint (IBVp)*

The aims of the Interaction Behaviour Viewpoint are shown in the Viewpoint Context View in Figure 8.14.

The main aim of the Interaction Behaviour Viewpoint is to 'Understand system life cycles' in terms of their interactions ('Understand interactions between life cycles') from a behavioural point of view ('Understand life cycle behaviour'), in a way that can 'Apply to any type of life cycle'.

#### 8.3.6.1    Description

The VDV in Figure 8.15 shows the Ontology Elements that appear on an Interaction Behaviour Viewpoint.

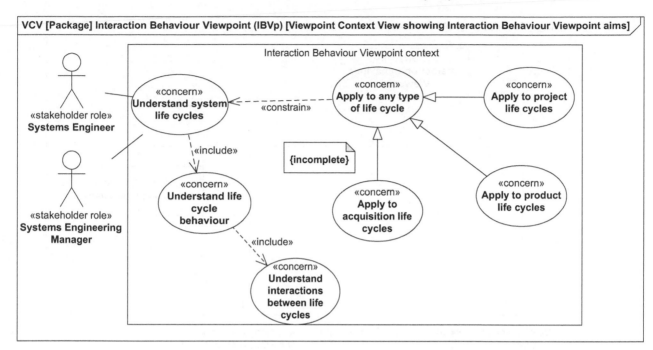

*Figure 8.14   Viewpoint Context View showing Interaction Behaviour Viewpoint aims*

The Interaction Behaviour Viewpoint shows one or more Life Cycle Models, the Stages that make them up and the Life Cycle Interactions between the Life Cycle Models. These Life Cycle Interactions must correspond to the Life Cycle Interaction Points between the Stages as defined on the corresponding Interaction Identification View. Note that the Interaction Behaviour View shows behavioural information – it shows how the Life Cycle Models interact through time.

### 8.3.6.2   Example Interaction Behaviour Viewpoint

Two example Views that conform to the Interaction Behaviour Viewpoint are shown in Figures 8.16 and 8.17.

Figure 8.16 shows an Interaction Behaviour View realised as a SysML *sequence diagram*. It shows the Life Cycle Interactions for a single Life Cycle Model (the 'Development Life Cycle'). The Stages of the Life Cycle are represented by the life lines. Each Life Cycle Interaction is represented by a *message* which enter and exit the *sequence diagram* via *gates*. (Note: these are SysML *gates*, part of the notation of the *sequence diagram*. They are not the same as the Ontology Element 'Gate' that appears on Figure 8.2). Note also that the full term Life Cycle

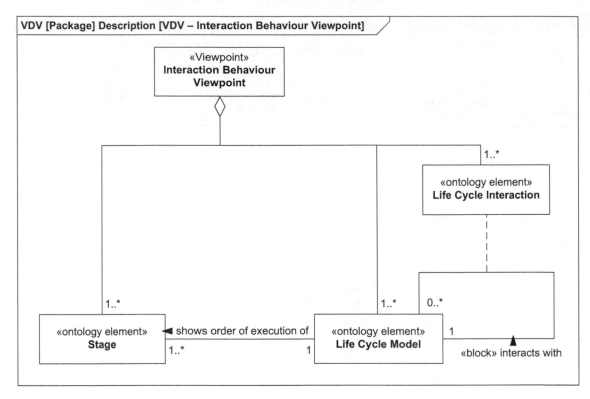

*Figure 8.15    Viewpoint Definition View showing the elements that appear on the
Interaction Behaviour Viewpoint (IBVp)*

Interaction is not used here for the stereotype on the *messages* purely for reasons of clarity and presentation.

With the visualisation used in Figure 8.16 it should be noted that the other Life Cycles that are interacted with are not shown, simply the Life Cycle Interaction Points entering and leaving via *gates*. If the interactions between the different Life Cycle Models are particularly important, then the alternate visualisation shown in Figure 8.17 may be considered.

Figure 8.17 has a similar visualisation to that shown in Figure 8.16 but this time the Life Cycle Interaction Points do not enter and exit the diagram anonymously, but interact with specific Life Cycles Models. Each Life Cycle Model is shown as a *boundary* with each Stage visualised as a *life line*. This time, however, the Life Cycle Interaction Points do not end in a *gate* but go to other Stages (the *life lines*) within other Life Cycle Models (the *packages*).

The Interaction Behaviour Views in Figures 8.16 and 8.17, along with the Interaction Identification View in Figure 8.13, fulfil Rule LC5.

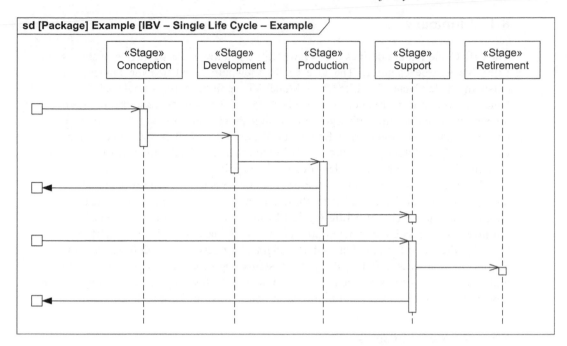

*Figure 8.16   An example of an Interaction Behaviour View*

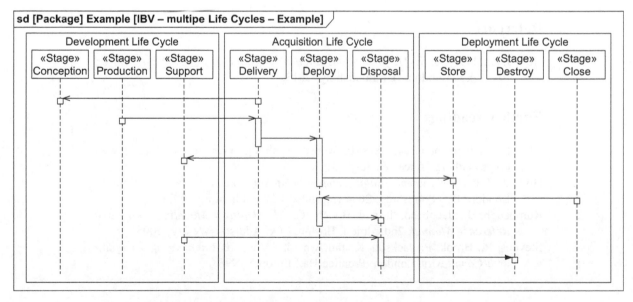

*Figure 8.17   An example of an Interaction Behaviour View showing full Life Cycle Interactions*

## 8.4    Summary

The Life Cycle Pattern defines four Viewpoints that allow the aspects of a System's Life Cycle to be captured. The Life Cycle Viewpoint identifies the Stages that exists in a Life Cycle. The Life Cycle Model Viewpoint describes how each Stage behaves over time in relation to the other Stages. The Interaction Identification Viewpoint identifies the Life Cycle Interaction Points between a number of Life Cycles. Finally, the Interaction Behaviour Viewpoint describes the behaviour of each Life Cycle Interaction Point in relation to other Life Cycle Interaction Points as identified in an Interaction Identification Viewpoint.

The Life Cycle Model Viewpoint presented here is the minimal version of the Viewpoint – it can be extended to show the Process Execution Groups that are executed in each Stage of a Life Cycle Model as well as the processes executed within each Process Execution Group. This allows the use of the Life Cycle Pattern to bridge the gap often found in Systems Engineering between the defined Systems Engineering processes followed by the systems engineers developing a System and the project plans produced by project managers tasked with managing the development of the System. For a full discussion of this, see Holt and Perry (2013).

## 8.5    Related Patterns

If using the Life Cycle pattern, the following patterns may also be of use:

- Epoch Pattern.

## Reference

Holt, J. and Perry, S. *SysML for Systems Engineering; 2nd Edition: A Model-Based Approach*. London: IET Publishing; 2013.

## Further readings

Object Management Group. SysML website [online]. Available at: http://www. omgsysml.org [Accessed July 2011].

Holt, J. *UML for Systems Engineering – Watching the Wheels*. 2nd Edition. London: IEE Publishing; 2004 (reprinted IET Publishing; 2007).

Rumbaugh, J., Jacobson, I. and Booch, G. *The Unified Modeling Language Reference Manual*. 2nd edition. Boston, MA: Addison-Wesley; 2005.

Stevens, R., Brook, P., Jackson, K. and Arnold, S. *Systems Engineering – Coping with Complexity*. London: Prentice Hall Europe; 1998.

*Chapter 9*

# Evidence Pattern

## 9.1 Introduction

The ability to present claims backed up by arguments that are supported by evidence, together with the ability to counter such claims in a similar manner, is a common need in systems engineering. While most often seen in areas such as testing, it applies equally to the establishment of traceability, definition of safety cases and even the presentation of business cases. Whatever the reason, such chains of evidence-arguments-claims should be presented in a consistent and rigorous manner.

### 9.1.1 Pattern aims

This Pattern is intended to be used as an aid to the definition chains of evidence and arguments used to support claims made about a subject. The main aims of this Pattern are shown in the Architectural Framework Context View (AFCV) in Figure 9.1.

The main purpose of the Evidence Pattern is to 'Support definition of evidence-based claims' made by a 'Claimant'. This includes the Use Cases:

- 'Allow claims to be made' – allow claims to be made by a Claimant about a subject.
- 'Allow supporting arguments to be established' – allow the arguments that support a claim to be established.
- 'Allow reinforcing evidence to be established' – allow the evidence that reinforces an argument to be established.
- 'Ensure linking relationships are established' – allow the linking relationships from evidence to arguments to claims to be clearly established.

The Use Case 'Support definition of evidence-based claims' is constrained by the need that the Evidence Pattern 'Must allow counter-claims to be made'. That is, it must allow counter-claims to be made against any aspect of a claim (the claim itself, the supporting arguments, the reinforcing evidence or the links between them).

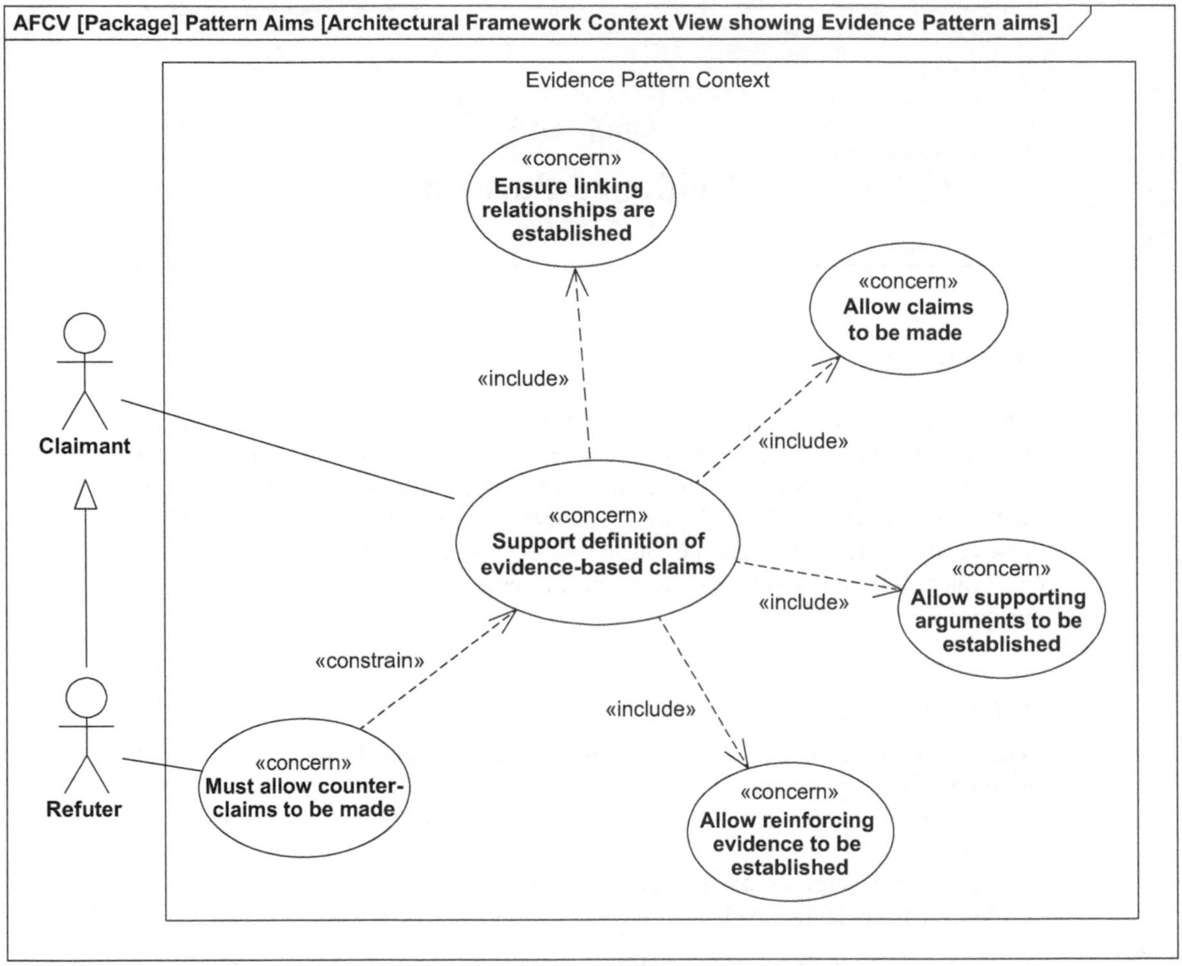

*Figure 9.1   Architectural Framework Context View showing Evidence Pattern aims*

## 9.2   Concepts

The main concepts covered by the Evidence Pattern are shown in the Ontology Definition View (ODV) in Figure 9.2.

Key to this Pattern is the concept of a 'Claim' that is made by a 'Claimant' about a 'Subject'. Such a 'Claim' is supported by one or more 'Argument' via a 'Claim-Argument Link'. Each 'Argument' is itself reinforced by one or more 'Evidence' via an 'Argument-Evidence Link'.

Note that a 'Claim', 'Argument', 'Claim-Argument Link', 'Evidence' and 'Argument-Evidence Link' are all types of the abstract concept of a 'Claimable Item'. This allows a 'Counter-Claim' (a special type of 'Claim') to be made about

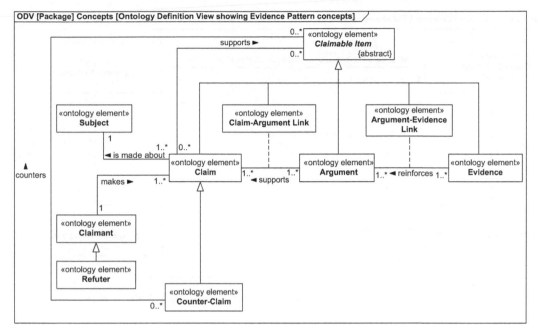

*Figure 9.2    Ontology Definition View showing Evidence Pattern concepts*

any type of 'Claimable Item' and a 'Claim' to support any type of 'Claimable Item'. Such a 'Counter-Claim' is made by a 'Refuter', a special type of 'Claimant'.

## 9.3    Viewpoints

This section describes the Viewpoints that make up the Evidence Pattern. It begins with an overview of the Viewpoints, defines Rules that apply to the Pattern and then defines each Viewpoint.

### 9.3.1    Overview

The Evidence Pattern defines a number of Viewpoints as shown in the Viewpoint Relationship View (VRV) in Figure 9.3.

The Evidence Pattern defines four Viewpoints for the definition of Evidence-Argument-Claim chains:

- The 'Claim Definition Viewpoint' is used to define Claims for a particular Subject and to show who made the Claims.
- The 'Argument Viewpoint' is used to show the Arguments that support a Claim.
- The 'Evidence Viewpoint' is used to show the Evidence that reinforces Arguments.
- The 'Counter-Claim Viewpoint' is used to make Counter-Claims (or supporting Claims) about Claimable Items.

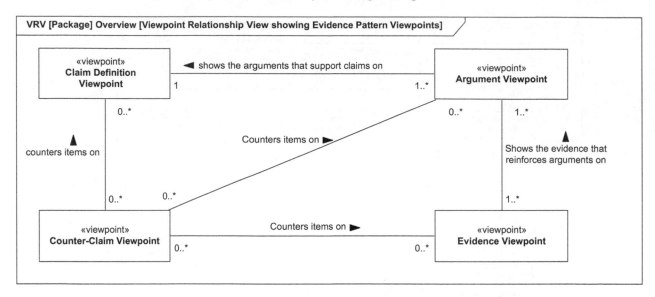

*Figure 9.3    Viewpoint Relationship View showing Evidence Pattern Viewpoints*

Each of these Viewpoints is described in more detail in the following sections. For each Viewpoint an example is also given.

### 9.3.2    Rules

Seven rules apply to the four Evidence Viewpoints, as shown in the Rules Definition View (RDV) in Figure 9.4.

These seven Rules are:

- 'Rule EP01: Every Claim must be supported by at least one Argument.' A Claim cannot be made without some supporting Argument.
- 'Rule EP02: Every Argument must be reinforced by at least one Evidence.' An Argument is only valid if there is some Evidence to back it up.
- 'Rule EP03: Every Claim must be made by a defined Claimant.' For a Claim to be valid, there must be a Claimant associated with it who has made the Claim. Anonymous Claims are not permitted.
- 'Rule EP04: Every Claim must be made about an identified Subject.' Claims cannot be generic, they have to be about something, that is they have to have a Subject.
- 'Rule EP05: As a minimum one Claim Definition View, one Argument View and one Evidence View must be produced.' This Rule establishes the minimum set of Views that have to be produced when using the Pattern for the use to be valid. If any are missing then all the necessary parts of the stated claim will not be present.

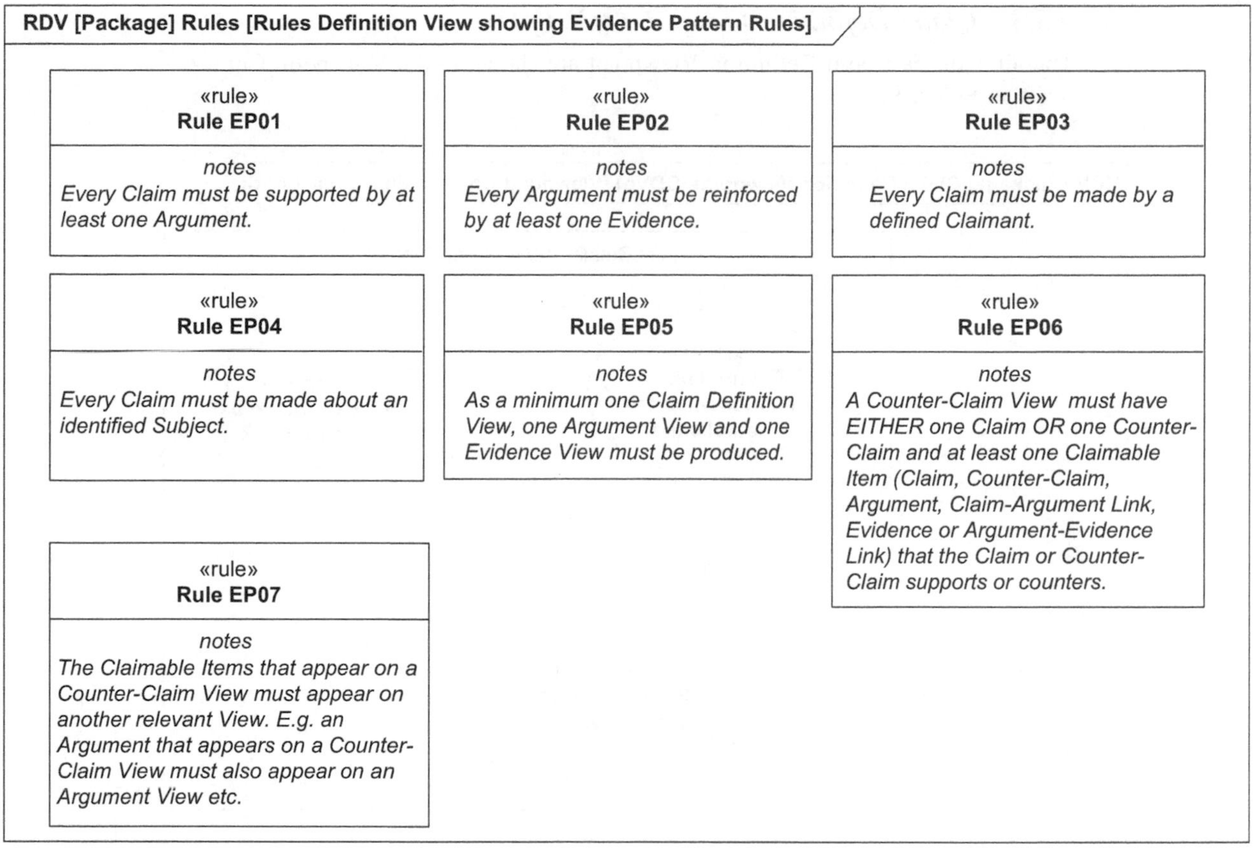

**RDV [Package] Rules [Rules Definition View showing Evidence Pattern Rules]**

| «rule»<br>**Rule EP01** | «rule»<br>**Rule EP02** | «rule»<br>**Rule EP03** |
|---|---|---|
| *notes*<br>*Every Claim must be supported by at least one Argument.* | *notes*<br>*Every Argument must be reinforced by at least one Evidence.* | *notes*<br>*Every Claim must be made by a defined Claimant.* |

| «rule»<br>**Rule EP04** | «rule»<br>**Rule EP05** | «rule»<br>**Rule EP06** |
|---|---|---|
| *notes*<br>*Every Claim must be made about an identified Subject.* | *notes*<br>*As a minimum one Claim Definition View, one Argument View and one Evidence View must be produced.* | *notes*<br>*A Counter-Claim View must have EITHER one Claim OR one Counter-Claim and at least one Claimable Item (Claim, Counter-Claim, Argument, Claim-Argument Link, Evidence or Argument-Evidence Link) that the Claim or Counter-Claim supports or counters.* |

«rule»<br>**Rule EP07**

*notes*
*The Claimable Items that appear on a Counter-Claim View must appear on another relevant View. E.g. an Argument that appears on a Counter-Claim View must also appear on an Argument View etc.*

*Figure 9.4   Rules Definition View showing Evidence Pattern Rules*

- 'Rule EP06: A Counter-Claim View must have EITHER one Claim OR one Counter-Claim and at least one Claimable Item (Claim, Counter-Claim, Argument, Claim-Argument Link, Evidence or Argument-Evidence Link) that the Claim or Counter-Claim supports or counters.' This Rule simply states that you cannot have an "empty" Counter-Claim View, i.e. one with no Claim in support of something else or Counter-Claim that refutes something else.
- 'Rule EP07: The Claimable Items that appear on a Counter-Claim View must appear on another relevant View. E.g. an Argument that appears on a Counter-Claim View must also appear on an Argument View etc.' This Rule simply states that if creating a Counter-Claim View, then that View must be making a Claim or Counter-Claim about something that has already been defined on one of the other Views; you can't counter or support something that hasn't already been claimed.

Note that the seven Rules shown in Figure 9.4 are the minimum that are needed. Others could be added if required.

### 9.3.3 *Claim Definition Viewpoint (CDVp)*

The aims of the Claim Definition Viewpoint are shown in the Viewpoint Context View in Figure 9.5.

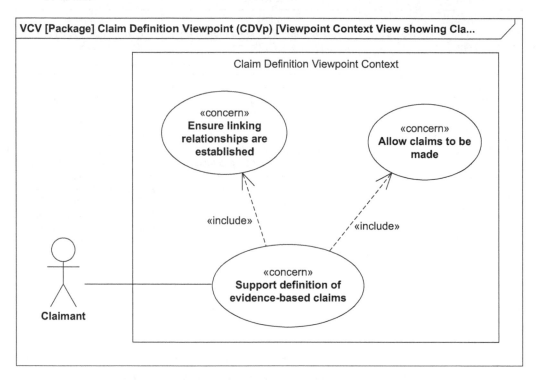

*Figure 9.5 Viewpoint Context View showing Claim Definition Viewpoint aims*

The main aim of the Claim Definition Viewpoint is to 'Support definition of evidence-based claims' through the aims of 'Allow claims to be made' and 'Ensure linking relationships are established'. That is, its aim is to capture the Claims made by a Claimant about a Subject together with the relationships from Claimant to Claim to Subject.

#### 9.3.3.1 Description

The Viewpoint Definition View (VDV) in Figure 9.6 shows the Ontology Elements that appear on a Claim Definition Viewpoint.

The Claim Definition Viewpoint identifies the Claims being made by a Claimant about a Subject.

As defined, the Claim Definition Viewpoint allows only a single Claimant and Subject to be shown, with any number of Claims made by that Claimant about the Subject. The definition could, of course, be extended to allow multiple Claimants and Subjects to be shown.

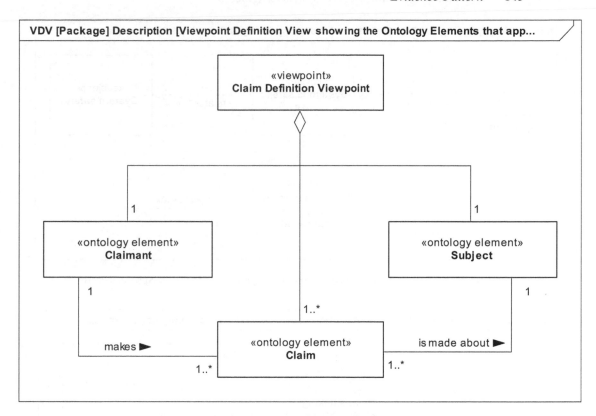

*Figure 9.6    Viewpoint Definition View showing the Ontology Elements that*
*appear on the Claim Definition Viewpoint (CDVp)*

### 9.3.3.2   Example

An example View that conforms to the Claim Definition Viewpoint is shown in Figure 9.7.

The Claim Definition View in Figure 9.7, realised as a SysML *block definition diagram*, shows the 'Safety Officer' Claimant making two Claims ('System is safe to use' and 'Safety requirements have been exceeded') about the Subject of 'System safety'.

The Claim and Subject are realised using stereotyped *blocks* and the Claimant using a stereotyped *actor*. Stereotyped *dependencies* have been used to realise the relationships between Claimant, Claims and Subject.

This View conforms to Rules EP03 and EP04. With Figures 9.10 and 9.13 it also conforms to Rule EP05.

### 9.3.4   *Argument Viewpoint (AVp)*

The aims of the Argument Viewpoint are shown in the Viewpoint Context View in Figure 9.8.

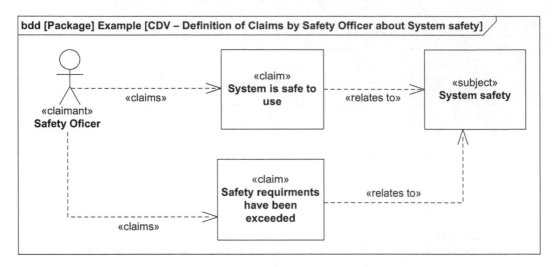

*Figure 9.7   CDV – Definition of Claims by Safety Officer about System safety*

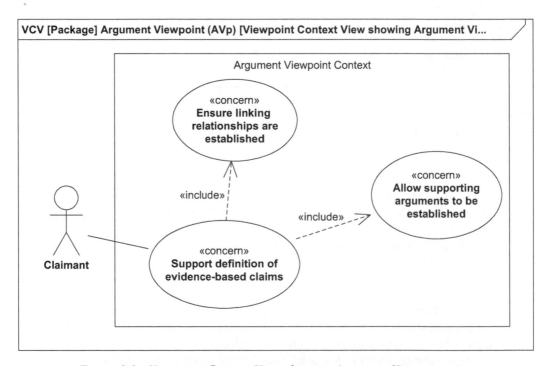

*Figure 9.8   Viewpoint Context View showing Argument Viewpoint aims*

The main aim of the Argument Viewpoint is to 'Support definition of evidence-based claims' through the aims of 'Allow supporting arguments to be established' and 'Ensure linking relationships are established'. That is, its aim is to capture the Arguments that support a Claim together with the relationships from the Arguments to the Claim.

### 9.3.4.1   Description

The VDV in Figure 9.9 shows the Ontology Elements that appear on an Argument Viewpoint.

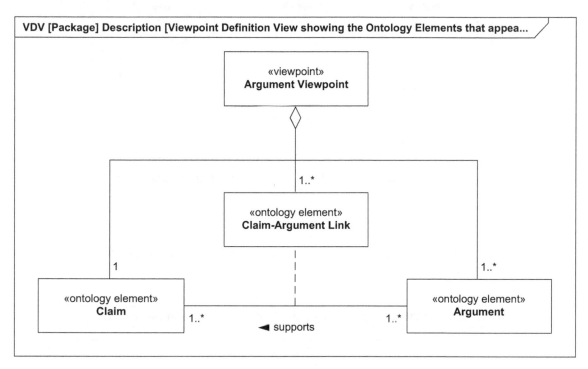

*Figure 9.9    Viewpoint Definition View showing the Ontology Elements that appear on the Argument Viewpoint (AVp)*

The Argument Viewpoint shows the Arguments that support a Claim.

As defined, the Argument Viewpoint only allows a single Claim to be shown. It could, of course, be extended to allow multiple Claims to be shown.

### 9.3.4.2   Example

An example View that conforms to the Argument Viewpoint is shown in Figure 9.10.

The Argument View in Figure 9.10, realised as a SysML *block definition diagram*, shows two Arguments ('System has been tested' and 'Safety statistics are good') that support the Claim that 'System is safe to use'.

A stereotyped *block* is used to realise the Arguments, with stereotyped *dependencies* realising the Claim-Argument Links.

This View conforms to Rule EP01. With Figures 9.7 and 9.13 it also conforms to Rule EP05.

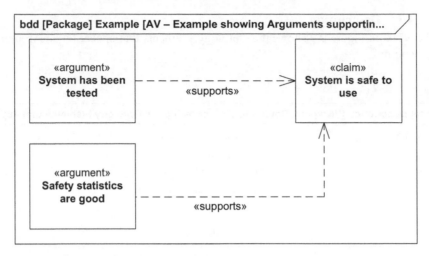

*Figure 9.10   AV – Example showing Arguments supporting "System is safe to use"
            Claim*

### 9.3.5   *Evidence Viewpoint (EVp)*

The aims of the Evidence Viewpoint are shown in the Viewpoint Context View in
Figure 9.11.

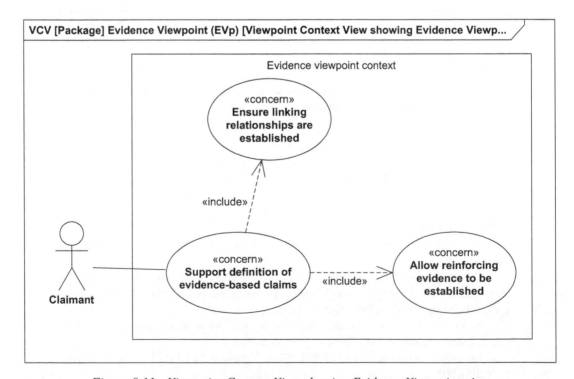

*Figure 9.11   Viewpoint Context View showing Evidence Viewpoint aims*

The main aim of the Evidence Viewpoint is to 'Support definition of evidence-based claims' through the aims of 'Allow reinforcing evidence to be established' and 'Ensure linking relationships are established'. That is, its aim is to capture the Evidence that reinforces an Argument together with the relationships from the Evidence to the Argument.

### 9.3.5.1    Description

The VDV in Figure 9.12 shows the Ontology Elements that appear on an Evidence Viewpoint.

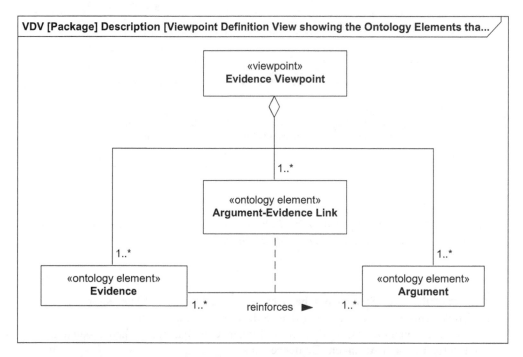

*Figure 9.12    Viewpoint Definition View showing the Ontology Elements that appear on the Evidence Viewpoint (EVp)*

The Evidence Viewpoint shows the Evidence that reinforces one or more Arguments.

### 9.3.5.2    Example

An example View that conforms to the Evidence Viewpoint is shown in Figure 9.13.

The Evidence View in Figure 9.13, realised as a SysML *block definition diagram*, shows three pieces of Evidence reinforcing two Arguments. The Two pieces of Evidence 'Safety case results' and 'Simulation results' both reinforce the

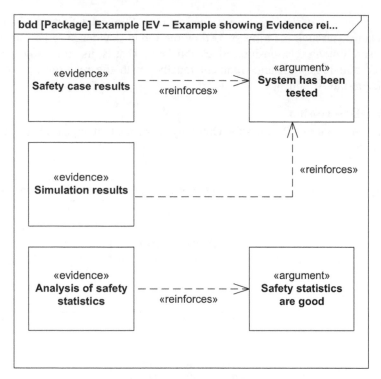

*Figure 9.13    EV – Example showing Evidence reinforcing the Arguments 'System has been tested' and 'Safety statistics are good'*

Argument that the 'System has been tested'. The Evidence 'Analysis of safety statistics' reinforces the Argument that 'Safety statistics are good'.

A stereotyped *block* is used to realise Evidence, with a stereotyped *dependency* used to realise each Argument-Evidence Link.

This View conforms to Rule EP02. With Figures 9.7 and 9.10 it also conforms to Rule EP05.

### 9.3.6    Counter-Claim Viewpoint (CCVp)

The aims of the Counter-Claim Viewpoint are shown in the Viewpoint Context View in Figure 9.14.

The main aim of the Counter-Claim Viewpoint is to 'Support definition of evidence-based claims' through the aims of 'Must allow counter-claims to be made' and 'Ensure linking relationships are established'. That is, its aim is to allow Counter-Claims to be made by a Refuter against any aspect of a claim (the Claim itself, the supporting Arguments, the reinforcing Evidence or the links between

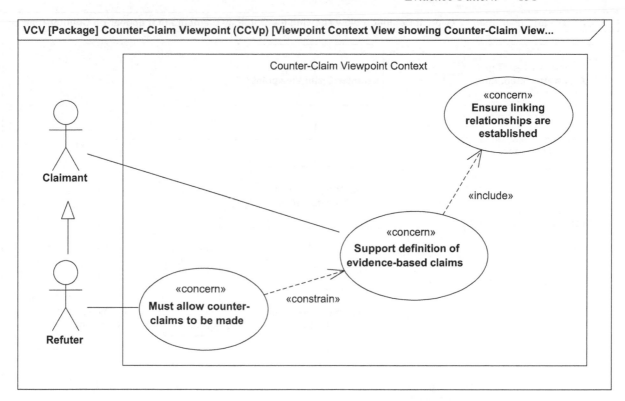

*Figure 9.14   Viewpoint Context View showing Counter-Claim Viewpoint aims*

them). Similarly it can be used by a Claimant to make a Claim in support of any aspect (the Claim itself, the supporting Arguments, the reinforcing Evidence or the links between them). The necessary links, for example from Refuter to Counter-Claim to item that the Counter-Claim counters, must also be captured.

### 9.3.6.1   Description

The VDV in Figure 9.15 shows the Ontology Elements that appear on a Counter-Claim Viewpoint.

The Counter-Claim Viewpoint is used to show a number of Counter-Claims and the Claimable Items that they counter OR a number of Claims and the Claimable Items that they support.

Although not made explicit in the VDV in Figure 9.15, only one Counter-Claim OR Claim can be shown on a CCV, as defined by Rule EP06. This rule also ensures that the Claimable Items that the Counter-Claims or Claims counter/support are also shown.

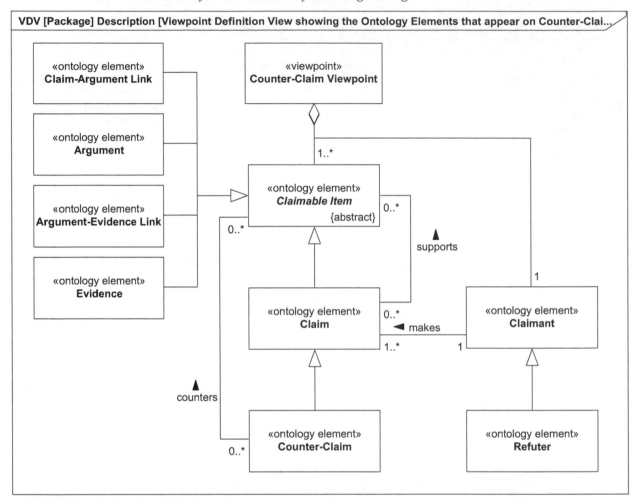

*Figure 9.15    Viewpoint Definition View showing the Ontology Elements that appear on Counter-Claim Viewpoint (CCVp)*

### 9.3.6.2  Example

An example View that conforms to the Counter-Claim Viewpoint is shown in Figure 9.16.

The Counter-Claim View in Figure 9.16, realised as a SysML *block definition diagram*, shows two pieces of Evidence ('Simulation results' and 'Safety case results') which reinforce, through the «reinforces» Argument-Evidence Links, the Argument that 'System has been tested'. These are all examples of Claimable Items.

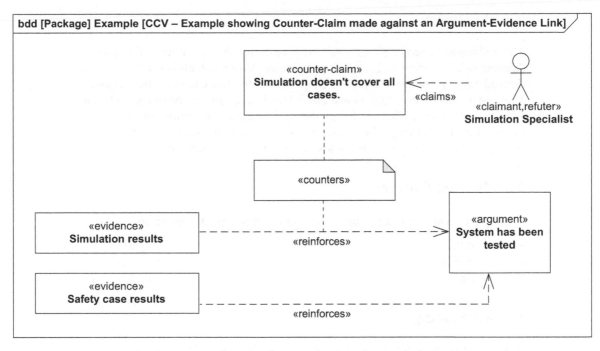

*Figure 9.16    CCV – Example showing Counter-Claim made against an
Argument-Evidence Link*

The diagram also shows a Counter-Claim that 'Simulation doesn't cover all cases' made by the Claimant 'Simulation Specialist' against the Argument-Evidence Link between the 'Simulation results' Evidence and the 'System has been tested' Argument.

As with the other Views, stereotyped *blocks* and *dependencies* have been used to realise the various Ontology Elements and their relationships, with the exception of the 'counters' relationship from the Counter-Claim to the Argument-Evidence Link. In this case a stereotyped *note* has been used which connects the Counter-Claim to the Argument-Evidence Link. This has been done because the SysML tool used to produce this diagram does not allow a *dependency* to target another *dependency*. For Counter-Claims made against Ontology Elements realised as SysML *blocks* (such as Evidence, Arguments), then a *dependency* stereotyped «counters» would be used. Note also the additional «refuter» stereotype applied to the *actor* representing the Claimant. This has been done to emphasise that this Claimant is playing a more specialised role of 'Refuter' (i.e. is a Claimant that is refuting something via a Counter-Claim).

This View conforms to Rule EP06. With Figure 9.13 it also conforms to Rule EP07.

## 9.4  Summary

The Evidence Pattern defines four Viewpoints for the definition of Evidence-Argument-Claim chains. The Claim Definition Viewpoint allows Claims to be defined for a particular Subject to show who made the Claims. The Argument Viewpoint allows the Arguments that support a Claim to be identified. The Evidence Viewpoint allows any supporting Evidence that reinforces Arguments to be identified. Finally, the Counter-Claim Viewpoint allows Counter-Claims (or supporting Claims) to be made about any type of Claimable Item.

## 9.5  Related Patterns

If using the Evidence Pattern, the following Patterns may also be of use:

- Description
- Traceability
- Testing.

## Further readings

Holt, J. and Perry, S. *SysML for Systems Engineering; 2nd Edition: A Model-Based Approach*. London: IET Publishing; 2013.

Object Management Group. *SysML website* [online]. Available at: http://www.omgsysml.org [Accessed February 2015].

Holt, J. *UML for Systems Engineering – Watching the Wheels*. 2nd Edition. London: IEE Publishing; 2004 (reprinted IET Publishing; 2007).

Rumbaugh, J., Jacobson, I. and Booch, G. *The Unified Modeling Language Reference Manual*. 2nd Edition. Boston, MA: Addison-Wesley; 2005.

*Chapter 10*

# Description Pattern

## 10.1  Introduction

When describing the elements of a system, there are a number of different aspects of an element's description that need to be considered. As well as the obvious aspects such as the element's properties and behaviours, it is also necessary to understand how an element relates to other elements. Such relationships are generally of three types:

- They may be general in nature, for example a relationship showing that a water supply supplies water to a tap.
- They may show how the element is broken down into parts, such as a tap being made up of a handle, spindle, valve mechanism.
- They may show "type" relationships between elements, for example showing that a mixer tap is a type of tap.

In addition, it is sometimes necessary to "localise" the descriptions of elements into different natural languages. Think, for example, of a set of installation or maintenance instructions for a mixer tap that may contain a description of the tap in multiple languages.

Without considering all of these aspects of an elements description, a full and consistent description may be lacking. The localisation aspect is particularly important in systems that are deployed, operated or maintained in different countries and therefore need multi-lingual yet consistent descriptions to be produced.

### 10.1.1  Pattern aims

This Pattern is intended to be used as an aid to the description of the elements of a system. The main aims of this Pattern are shown in the Architectural Framework Context View (AFCV) in Figure 10.1.

The key aim of the Description Pattern is to 'Support description of elements'. This includes the Use Cases:

- 'Support identification of properties' – the Pattern must support the identification of properties of an element, including any type information and initial or default property values.

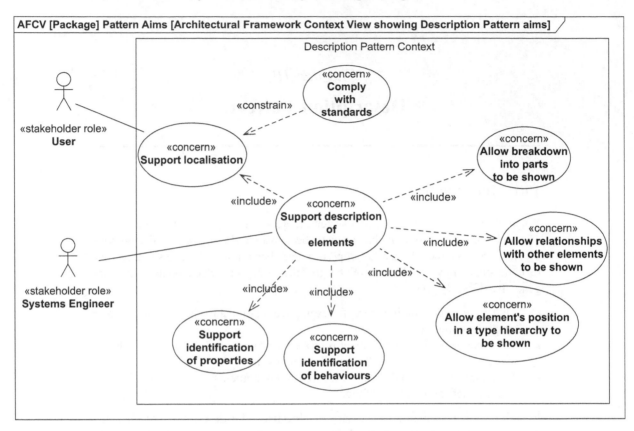

*Figure 10.1    Architectural Framework Context View showing Description Pattern aims*

- 'Support identification of behaviours' – the Pattern must support the identification of any behaviour that the element has. If necessary, parameterised behaviours must be able to be identified. Note that this Pattern is concerned purely with the identification of behaviour and not with its description, i.e. it identifies the behaviours that an element has but does not describe how the behaviours are carried out.
- 'Allow breakdown into parts to be shown' – the Pattern must allow the structural breakdown of an element into constituent parts to be shown.
- 'Allow element's position in a type hierarchy to be shown' – the Pattern must allow the element's position in a "type hierarchy" (often called a taxonomy) to be shown, allowing its relationship to more general types and more specialised types of the element to be shown.
- 'Allow relationships with other elements to be shown' – allow more general relationships (i.e. relationships other than through decomposition and taxonomy) to be shown.
- 'Support localisation' – the Pattern must allow extended textual descriptions of the element to be produced in a number of languages.

The Use Case 'Support localisation' is constrained by the need to 'Comply with standards'; it is important that the identification of languages used in localisation are clearly identified in conformance with best practice.

## 10.2   Concepts

The main concepts covered by the Description Pattern are shown in the Ontology Definition View (ODV) in Figure 10.2.

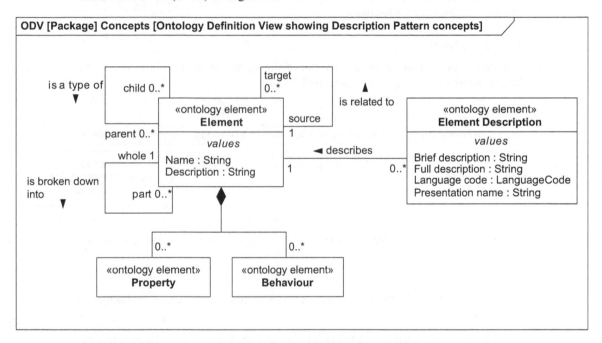

*Figure 10.2   Ontology Definition View showing Description Pattern concepts*

Key to this Pattern is the concept of an 'Element' that, at its most basic, has a 'Name' and a 'Description'. An 'Element' may also be composed of zero or more 'Property' and zero or more 'Behaviour'.

An 'Element' can be related to zero or more other target 'Element', may be broken down into zero or more 'Element' acting as parts of the whole and may be a type of zero or more parent 'Element'.

An 'Element' may be "localised" through further description by zero or more 'Element Description'. Each 'Element Description' has a:

- 'Presentation name' – the name used when referring to the localised 'Element'
- 'Brief description' – a short description of the 'Element'
- 'Full description' – a longer, more detailed, description of the 'Element'
- 'Language code' – a short code that identifies the language used in the 'Presentation name', 'Brief description' and 'Full description'

The localisation of an 'Element' through zero or more 'Element Description' is heavily constrained by the Rules associated with this Pattern. See the RDV for details.

## 10.3    Viewpoints

This section describes the Viewpoints that make up the Description Pattern. It begins with an overview of the Viewpoints, defines Rules that apply to the Pattern and then defines each Viewpoint.

### 10.3.1    Overview

The Description Pattern defines a number of Viewpoints as shown in the Viewpoint Relationship View (VRV) in Figure 10.3.

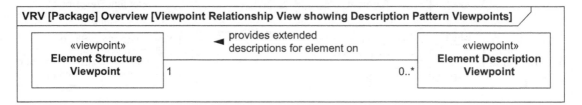

*Figure 10.3    Viewpoint Relationship View showing Description Pattern Viewpoints*

The Description Pattern defines two Viewpoints for the description of Elements:

- The 'Element Structure Viewpoint' is used to describe an Element in term of its Properties and Behaviours, relationship to other Elements (including relationships in a type taxonomy) and breakdown into parts.
- The 'Element Description Viewpoint' is used to show extended Element Descriptions for an Element on an 'Element Structure View'.

Each of these Viewpoints is described in more detail in the following sections. For each Viewpoint an example is also given.

### 10.3.2    Rules

Eight Rules apply to the two Description Viewpoints, as shown in the Rules Definition View (RDV) in Figure 10.4.
    The eight Rules are:

- 'Rule DP01: When using the Description Pattern, at least one Element Structure View must exist.' This Rule establishes the minimum set of Views that must be produced for the use of the Pattern to be valid.

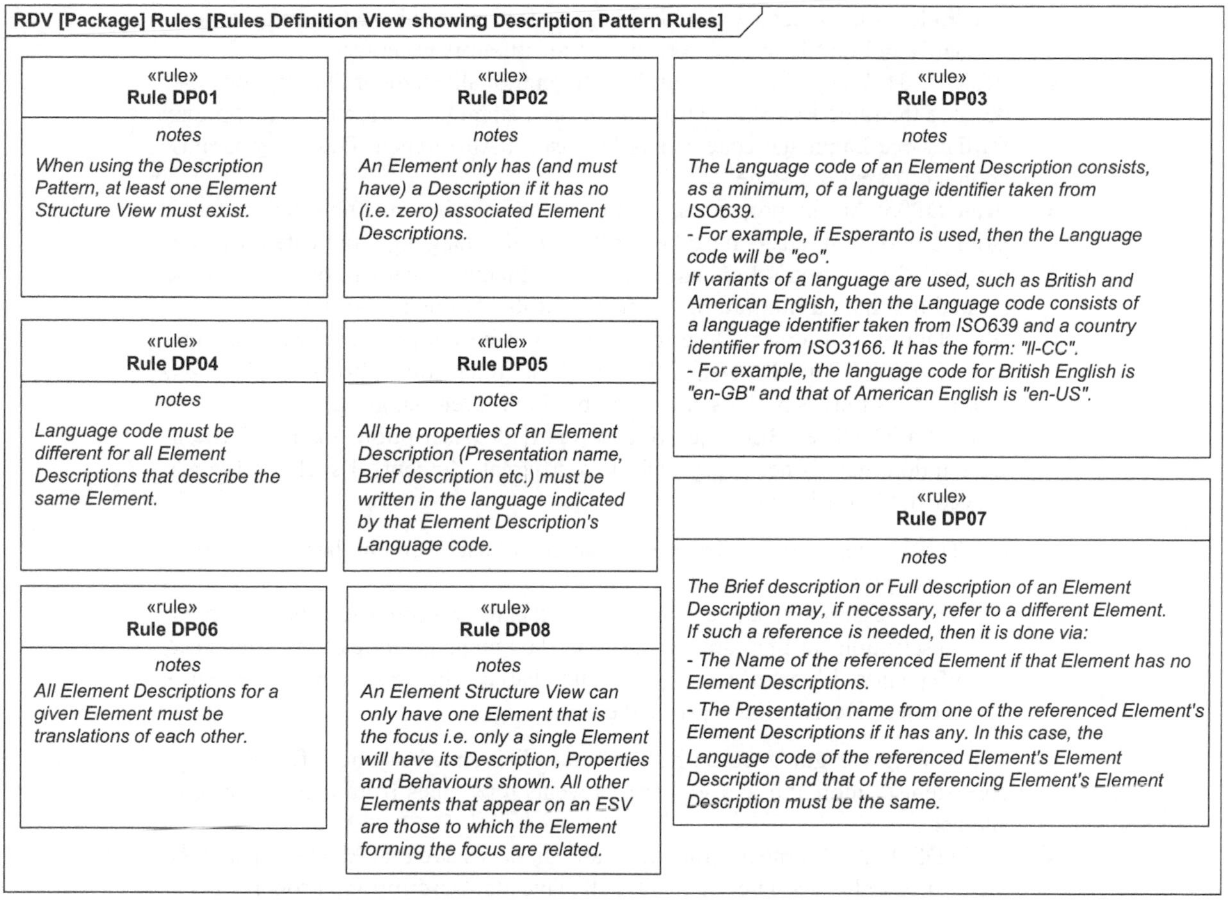

RDV [Package] Rules [Rules Definition View showing Description Pattern Rules]

| «rule» **Rule DP01** |
| --- |
| *notes* |
| *When using the Description Pattern, at least one Element Structure View must exist.* |

| «rule» **Rule DP02** |
| --- |
| *notes* |
| *An Element only has (and must have) a Description if it has no (i.e. zero) associated Element Descriptions.* |

| «rule» **Rule DP03** |
| --- |
| *notes* |
| *The Language code of an Element Description consists, as a minimum, of a language identifier taken from ISO639.*<br>*- For example, if Esperanto is used, then the Language code will be "eo".*<br>*If variants of a language are used, such as British and American English, then the Language code consists of a language identifier taken from ISO639 and a country identifier from ISO3166. It has the form: "ll-CC".*<br>*- For example, the language code for British English is "en-GB" and that of American English is "en-US".* |

| «rule» **Rule DP04** |
| --- |
| *notes* |
| *Language code must be different for all Element Descriptions that describe the same Element.* |

| «rule» **Rule DP05** |
| --- |
| *notes* |
| *All the properties of an Element Description (Presentation name, Brief description etc.) must be written in the language indicated by that Element Description's Language code.* |

| «rule» **Rule DP07** |
| --- |
| *notes* |
| *The Brief description or Full description of an Element Description may, if necessary, refer to a different Element. If such a reference is needed, then it is done via:*<br>*- The Name of the referenced Element if that Element has no Element Descriptions.*<br>*- The Presentation name from one of the referenced Element's Element Descriptions if it has any. In this case, the Language code of the referenced Element's Element Description and that of the referencing Element's Element Description must be the same.* |

| «rule» **Rule DP06** |
| --- |
| *notes* |
| *All Element Descriptions for a given Element must be translations of each other.* |

| «rule» **Rule DP08** |
| --- |
| *notes* |
| *An Element Structure View can only have one Element that is the focus i.e. only a single Element will have its Description, Properties and Behaviours shown. All other Elements that appear on an ESV are those to which the Element forming the focus are related.* |

*Figure 10.4   Rules Definition View showing Description Pattern Rules*

- 'Rule DP02: An Element only has (and must have) a Description if it has no (i.e. zero) associated Element Descriptions.' An Element's Description property is only completed (and must be so completed) if it has no associated Element Description. If it does have an Element Description then the Description property is not used.
- 'Rule DP03: The Language code of an Element Description consists, as a minimum, of a language identifier taken from ISO639.

    For example, if Esperanto is used then the Language code will be "eo."

  If variants of a language are used, such as British and American English, then the Language code consists of a language identifier taken from ISO639 and a country identifier from ISO3166. It has the form: "ll-CC."

    For example, the language code for British English is "en-GB" and that of American English is "en-US."

This Rule ensures that standard language codes are used based on international standards and that language variants are identified if necessary.

- 'Rule DP04: Language code must be different for all Element Descriptions that describe the same Element.' This Rule means that multiple Element Descriptions for the same Language code cannot be created. All Element Descriptions must be in a different language.
- 'Rule DP05: All the properties of an Element Description (Presentation name, Brief description etc.) must be written in the language indicated by that Element Description's Language code.' An Element Description must not be created that uses a mixture of languages for its various properties.
- 'Rule DP06: All Element Descriptions for a given Element must be translations of each other.' This Rule says that, as far as translation allows, all Element Descriptions must say the same thing, but in different languages.
- 'Rule DP07: The Brief description or Full description of an Element Description may, if necessary, refer to a different Element. If such a reference is needed, then it is done via:

  - The Name of the referenced Element if that Element has no Element Descriptions.
  - The Presentation name from one of the referenced Element's Element Descriptions if it has any. In this case, the Language code of the referenced Element's Element Description and that of the referencing Element's Element Description must be the same.'

  This Rule gives guidance on how different Elements may be referenced in a consistent fashion and in a way that prevents references from one language to another.
- 'Rule DP08: An Element Structure View can only have one Element that is the focus, i.e. only a single Element will have its Description, Properties and Behaviours shown. All other Elements that appear on an ESV are those to which the Element forming the focus are related.' This Rule simply states that only a single Element can be broken down, described or related on an ESV.

Note that the eight Rules shown in Figure 10.4 are the minimum that are needed. Others could be added if required.

## 10.3.3   Element Structure Viewpoint (ESVp)

The aims of the Element Structure Viewpoint are shown in the Viewpoint Context View in Figure 10.5.

The main aims of the Element Structure Viewpoint is to 'Support description of elements' through the aims of 'Support identification of properties', 'Support identification of behaviours', 'Allow breakdown into parts to be shown', 'Allow element's position in a type hierarchy to be shown' and 'Allow relationships with other elements to be shown'. That is, its aim is to describe an Element in term of its Properties and Behaviours, relationship to other Elements (including relationships in a type taxonomy) and breakdown into parts.

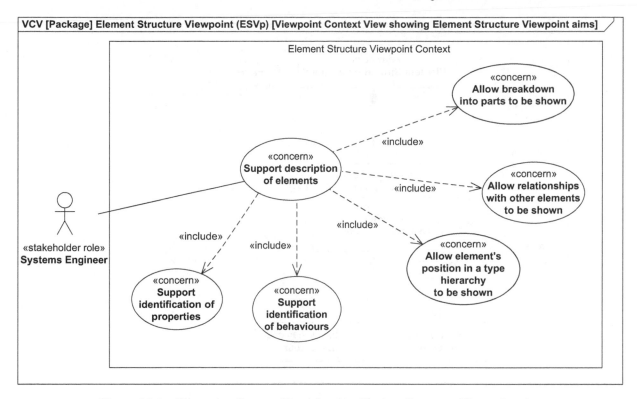

*Figure 10.5    Viewpoint Context View showing Element Structure Viewpoint aims*

### 10.3.3.1    Description

The Viewpoint Definition View (VDV) in Figure 10.6 shows the Ontology Elements that appear on an Element Structure Viewpoint.

The Element Structure Viewpoint is used to describe an Element in terms of its Properties and Behaviours, relationship to other Elements (including relationships in a type taxonomy) and breakdown into Elements representing its parts.

Although an Element Structure Viewpoint can show multiple Elements, it allows only a single Element to be the focus, i.e. only a single Element will have its Description, Properties and Behaviours shown. All other Elements that appear on a View which realises an Element Structure Viewpoint are those to which the Element forming the focus are related. The definition could, of course, be extended to allow multiple Elements to be shown in detail.

### 10.3.3.2    Example

An example View that conforms to the Element Structure Viewpoint is shown in Figure 10.7.

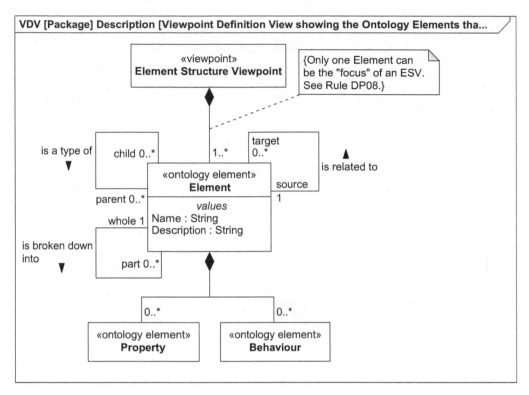

*Figure 10.6    Viewpoint Definition View showing the Ontology Elements that appear on the Element Structure Viewpoint (ESVp)*

The Element Structure View in Figure 10.7, realised as a SysML *block definition diagram*, shows the structure of the 'Water Tap' Element (highlighted in light grey for emphasis). Apologies to any plumbers who may be reading!

SysML *blocks* have been used to realise Elements, with the addition of a *stereotype* for emphasis. The 'Name' of each Element is represented by the name of the realising *block*.

The relationships between the 'Water Tap' and the other Elements are shown. Examples of all three types of relationship from the ODV are shown: 'is a type of' is realised using *generalisation*, 'is broken down into' is realised by *composition* and 'is related to' through *association*.

The Properties of the 'Water Tap' Element are shown as SysML *properties* and the Behaviour of the 'Water Tap' through SysML *operations*.

This diagram conforms to Rule DP01. Note that 'Water Tap' has no Description; in accordance with Rule DP02 this means that 'Water Tap' must have at least one Element Description associated with it. This has been done and is shown in Figure 10.10.

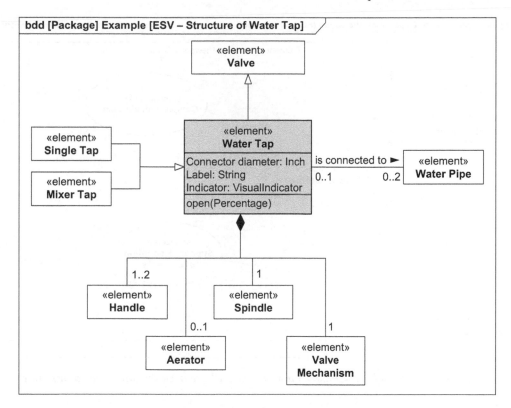

*Figure 10.7   ESV – Structure of Water Tap*

## 10.3.4   Element Description Viewpoint (EDVp)

The aims of the Element Description Viewpoint are shown in the Viewpoint Context View in Figure 10.8.

The key aim of the Element Description Viewpoint is to 'Support description of elements' through the aim of 'Support localisation' while meeting the constraint of 'Comply with standards'. That is, the Viewpoint is concerned with supporting the production of localised textual descriptions of an Element.

### 10.3.4.1   Description

The VDV in Figure 10.9 shows the Ontology Elements that appear on an Element Description Viewpoint.

The Element Description Viewpoint is used to show extended Element Descriptions for a single Element.

As defined, the Element Description Viewpoint allows only a single Element to be the focus, i.e. only a single Element will have its extended descriptions in terms of Element Descriptions shown. The definition could, of course, be extended to allow multiple Elements to be shown in detail.

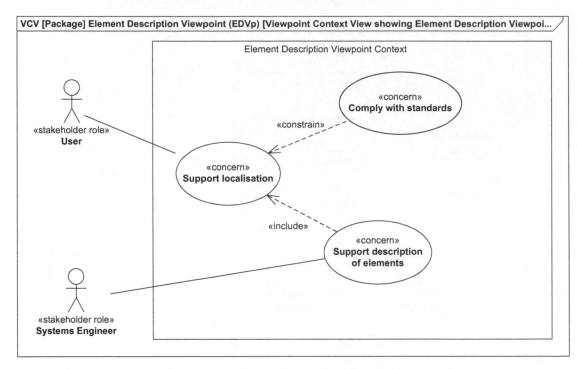

*Figure 10.8    Viewpoint Context View showing Element Description Viewpoint aims*

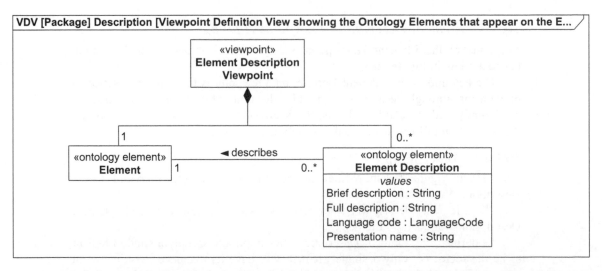

*Figure 10.9    Viewpoint Definition View showing the Ontology Elements that appear on the Element Description Viewpoint (EDVp)*

### 10.3.4.2   Example

An example View that conforms to the Element Description Viewpoint is shown in Figure 10.10.

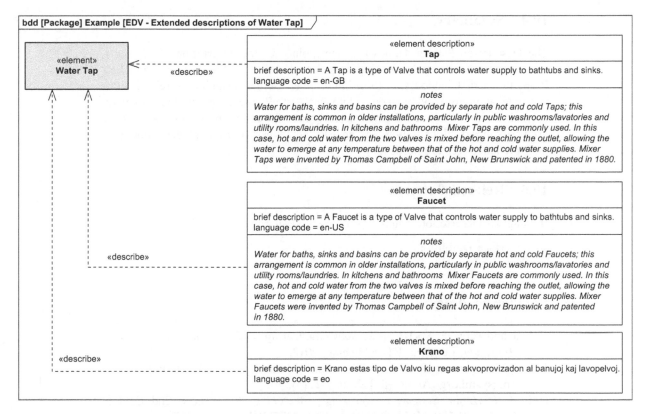

*Figure 10.10   EDV – Extended descriptions of Water Tap*

The Element Description View in Figure 10.10, realised as a SysML *block definition diagram*, shows the localisation through extended Element Descriptions of the 'Water Tap' Element (highlighted in light grey for emphasis).

Each Element Description is realised as a SysML *block* and has a *stereotype* added for extra emphasis. The 'Presentation name' of each Element Description is represented by the name of the realising *block*. The 'Brief description' and 'Language code' are realised using *tags* associated with the *stereotype*.

The 'Extended description' of each Element Description is realised using the "notes" field of the realising *block*. Note that while "notes" is not a built-in SysML concept, all SysML modelling tools allow notes or descriptive text to be added against elements in the tool. The 'Extended description' has been omitted for brevity for the Element Description written in Esperanto (the Element Description with 'language code' set to 'eo').

The relationship between the Element Descriptions and the Elements that they describe are realised using stereotyped *dependencies*.

This diagram conforms to Rules DP02 to DP07.

## 10.4  Summary

The Description Pattern defines two Viewpoints for the description of Elements. The Element Structure Viewpoint is used to describe an Element in term of its Properties and Behaviours, relationship to other Elements (including relationships in a type taxonomy) and breakdown into parts. The Element Description Viewpoint is used to show extended Element Descriptions for an Element; this can also be used to support "localisation" of an Element, where consistent descriptions in multiple languages are needed.

## 10.5  Related Patterns

If using the Description Pattern, the following Patterns may also be of use:

- Interface Definition
- Traceability.

## Further readings

Holt, J. and Perry, S. *SysML for Systems Engineering; 2nd Edition: A Model-Based Approach.* London: IET Publishing; 2013.

Object Management Group. *SysML website* [online]. Available at: http://www.omgsysml.org [Accessed: February 2015].

Holt, J. *UML for Systems Engineering – Watching the Wheels.* 2nd Edition. London: IEE Publishing; 2004 (reprinted IET Publishing; 2007).

Rumbaugh, J., Jacobson, I. and Booch, G. *The Unified Modeling Language Reference Manual.* 2nd edition. Boston, MA: Addison-Wesley; 2005.

ISO 639 Overview. Available at: http://en.wikipedia.org/wiki/ISO_639 [Accessed: October 2015].

ISO 639-1 Codes: Available at: http://en.wikipedia.org/wiki/List_of_ISO_639-1_codes [Accessed: October 2015].

Country Codes. Available at: http://en.wikipedia.org/wiki/ISO_3166 [Accessed: October 2015].

Internet Engineering Task Force – Best Current Practice (BCP) 47 – Language Tags. Available from http://tools.ietf.org/html/bcp47 [Accessed: October 2015.]

*Chapter 11*

# Context Pattern

## 11.1 Introduction

Context is the basis for any piece of work, without context we cannot understand what we are working towards or what the extent of the work is. Context provides the reference to enable comparisons to be made between outcomes.

### 11.1.1 Pattern aims

This pattern is used to define a specific context (point of view) of a model that represents an aspect of that model from that point of view, such as stakeholder, system, project, life cycle stage, The main aims of this pattern are shown in the Architectural Framework Context View (ACFV) in Figure 11.1.

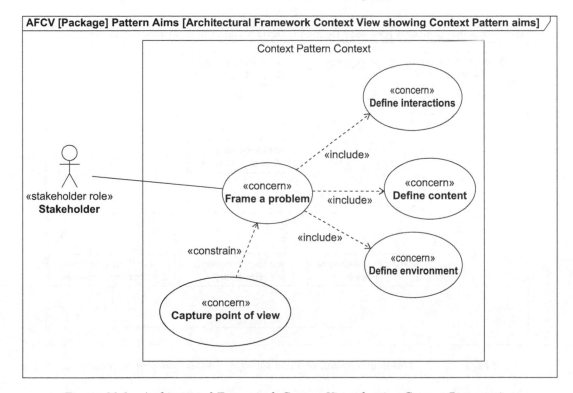

*Figure 11.1   Architectural Framework Context View showing Context Pattern aims*

The key aim of the context pattern is to 'Frame a problem' this helps to provide an understanding of a how something fits into the world around it. 'Frame a problem' includes the Use Cases:

- 'Define environment' – the environment is the world immediately outside the problem being framed, this is about understanding and recording those environmental elements.
- 'Define content' – the content is often the boundary of the system, but may also be those things inside the system those things which need to be done to solve the problem.
- 'Define interactions' – this recognises the relationships between the system and the world around it where the content defines something about the internals of the system there may also be interactions between the elements within.

All of these Use Cases are constrained by 'Capture point of view' focusing on the specific area or view by which a stakeholder or a system needs to be understood.

## 11.2   Concepts

The main concepts covered by the Context Pattern are shown in the Ontology Definition View (ODV) in Figure 11.2.

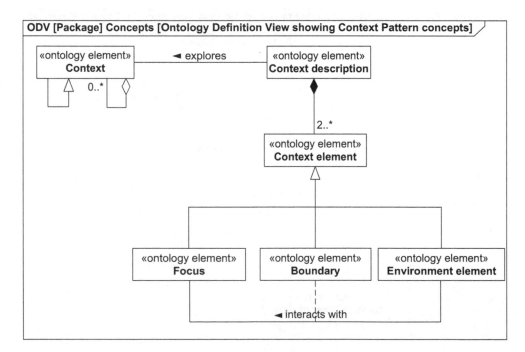

*Figure 11.2   Ontology Definition View showing Context Pattern concepts*

Key to this pattern is the concept of 'context'. A 'context' is a way of framing a problem by exploring a point of view often this is how it fits into the world around it. A 'Context' is explored through the use of a 'Context description' which enables the visualisation of the 'Context'. A 'Context description' is made up of at least two 'Context element'. 'Context element' has three types 'Focus', 'Boundary' and 'Environment element'.

The 'Environment element' represents those things in the outside world which are affected by the 'Context'.

The 'Focus' represents those things within the system which are to be considered, these could be concerns, needs, requirements or represents physical aspects to be considered.

The 'Boundary' defines the point at which the 'Focus' and 'Environment element' touch or interact.

## 11.3   Viewpoints

This section describes the Viewpoints that make up the Context Pattern. It begins with an overview of the Viewpoints, defines Rules that apply to the Pattern and then defines each Viewpoint.

### 11.3.1   Overview

The Context Pattern defines two Viewpoints as shown in the Viewpoint Relationship View (VRV) in Figure 11.3.

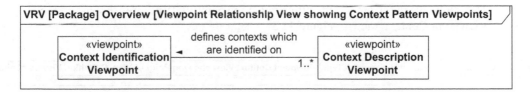

*Figure 11.3   Viewpoint Relationship View showing Context Pattern Viewpoints*

The Context Pattern defines two Viewpoints for the definition of Contexts:

- The 'Context Identification Viewpoint' is used to define the list of 'Context' to be considered on one or more 'Context Description Viewpoint'.
- The 'Context Description Viewpoint' is used to explore the contexts including 'Focus', 'Boundary' and 'Environment element'.

Each of these Viewpoints is described in more detail in the following sections. For each Viewpoint an example is also given.

### 11.3.2   Rules

Four rules apply to the two Context Viewpoints, as shown in the Rules Definition View (RDV) in Figure 11.4.

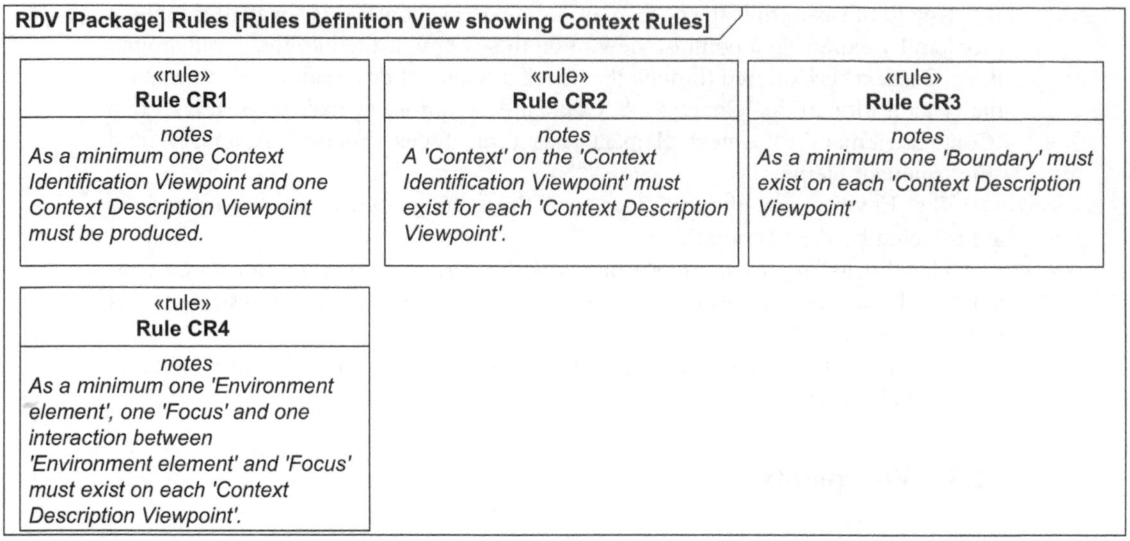

RDV [Package] Rules [Rules Definition View showing Context Rules]

«rule»
**Rule CR1**

*notes*
*As a minimum one Context Identification Viewpoint and one Context Description Viewpoint must be produced.*

«rule»
**Rule CR2**

*notes*
*A 'Context' on the 'Context Identification Viewpoint' must exist for each 'Context Description Viewpoint'.*

«rule»
**Rule CR3**

*notes*
*As a minimum one 'Boundary' must exist on each 'Context Description Viewpoint'*

«rule»
**Rule CR4**

*notes*
*As a minimum one 'Environment element', one 'Focus' and one interaction between 'Environment element' and 'Focus' must exist on each 'Context Description Viewpoint'.*

*Figure 11.4    Rules Definition View showing Context Rules*

Note that the four Rules shown in Figure 11.4 are the minimum that are needed. Others could be added if required.

## 11.3.3    Context Identification Viewpoint (CIVp)

The aims of the Context Identification Viewpoint (CIVp) are shown in the Viewpoint Context View in Figure 11.5.

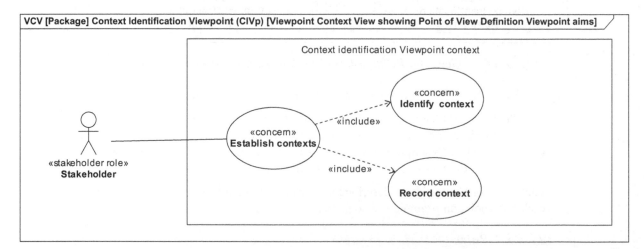

*Figure 11.5    Viewpoint Context View showing Point of View Definition Viewpoint aims*

The main aim of the CIVp is to 'Establish contexts' this is to understand the 'Context' to be explored. This includes the Use Cases 'Identify context' recognising the views which need to be considered and 'Record context' to ensure the focus of the context is captured.

### 11.3.3.1   Description

The Viewpoint Definition View (VDV) in Figure 11.6 shows the Ontology Elements that appear on a CIVp.

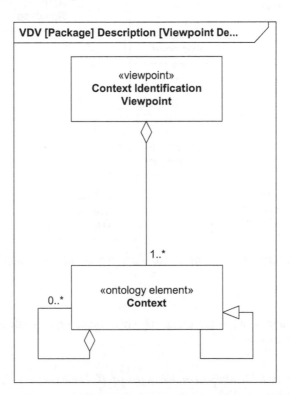

*Figure 11.6    Viewpoint Definition View showing the Ontology Elements that appear on the Context Identification Viewpoint (CIVp)*

The CIVp shows the contexts and the relationships which can be used to connect them together.

Many 'Context' can be shown on this viewpoint, any relationships between the 'context' can also be shown either by types representing the stakeholder needs or aggregating contexts to show the level of abstraction in a hierarchy.

### 11.3.3.2   Example Context Identification Viewpoint

An Example View that conforms to the CIVp is shown in Figure 11.7.

In the example in Figure 11.7 a taxonomy of contexts is described showing the types of stakeholder which have an interest in a system. It also shows a 'Railway'

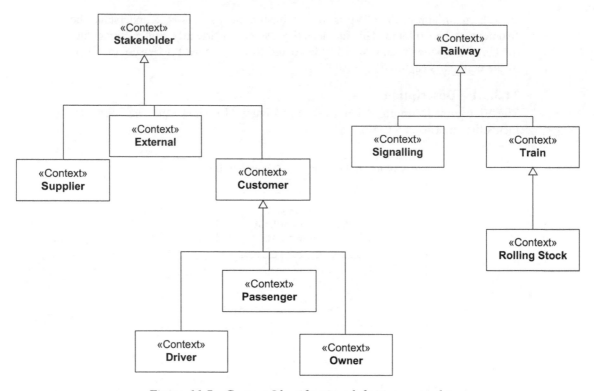

*Figure 11.7    Context Identification definition example*

as a 'Context' and some of the elements which make up a 'Railway' and require 'Context' to be described.

## 11.3.4    Context Description Viewpoint (CDVp)

The aims of the CDVp are shown in the Viewpoint Context View in Figure 11.8.
The main aim of the CDVp is to 'Define context', which includes the Use Cases:

- 'Define environment' – where the elements outside the system are defined,
- 'Define content' – where the system itself and relevant internal information is defined, and
- 'Define interactions' – where relationships between the environment and the content are defined.

The Systems Engineer and stakeholder are both affected by 'Define context' the stakeholder as it helps them to recognise what they expect and the Systems Engineer records and orders the context.

### 11.3.4.1    Description

The VDV in Figure 11.9 shows the Ontology Elements that appear on a Context view Viewpoint.

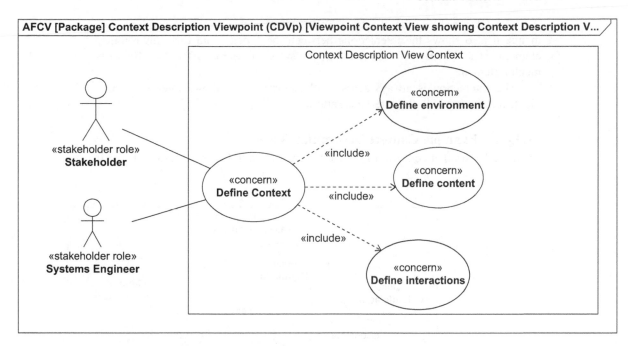

*Figure 11.8    Viewpoint Context View showing Context Description*
*Viewpoint aims*

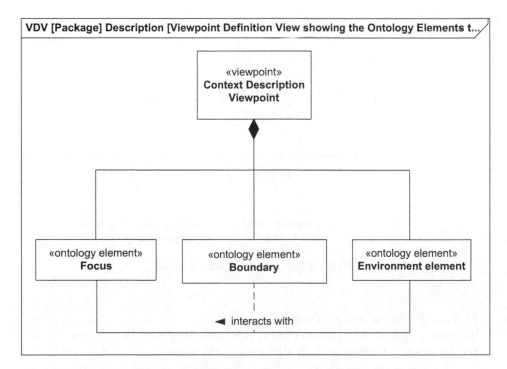

*Figure 11.9    Viewpoint Definition View showing the Ontology Elements that*
*appear on the Context Description Viewpoint (CDVp)*

The CDVp shows the 'Focus', 'Environment element' and 'Boundary' elements, it also shows the interactions between the 'Focus' and the 'Environment element'. The 'Focus' defines the things the system does that affect the 'Environmental elements'.

The interaction is always across a 'Boundary', this allows interfaces to be identified linking with the Interface Pattern.

### 11.3.4.2 Example Context Description View
Example Views that conform to the CDVp are shown in Figures 11.10 and 11.11.

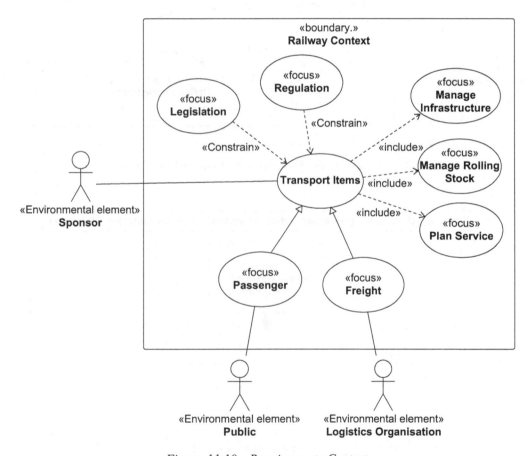

*Figure 11.10 Requirements Context*

#### 11.3.4.2.1 Requirements Context
Requirements need to be put into context to be considered useful to a system as the context captures the point of view they are defined from.

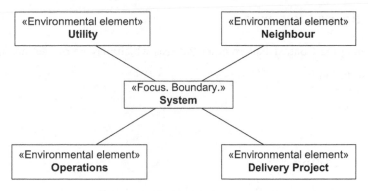

*Figure 11.11   System Context*

The example context in Figure 11.10 shows both the 'Environmental elements', what the system interacts with and the 'Focus' of the system what the system needs to achieve.

In Figure 11.10 the actors around the outside of the diagram define the environmental elements, the box defines the boundary and the use cases represent the content which is being put into context. Finally the solid line between actor and use case shows that an interaction exists between the environment and content.

The example is an extract from the requirements for a Railway where a 'Sponsor' is interested in 'Transport Items' there are two types of 'Transport Item' 'Passenger', which the 'Public' has an interaction with and 'Freight' which 'Logistics Organisation' have an interest in. To 'Transport items' includes 'Manage Infrastructure', 'Manage Rolling Stock' and 'Plan Service'. All of this must be delivered within the constraints of 'Legislation' and 'Regulation'.

### 11.3.4.2.2   System Context

In the example shown in Figure 11.11 the 'Boundary' is synonymous with the 'Focus' for the System. This is often the case when the interest is on the 'environmental elements' those things outside the system which interact with it.

Figure 11.11 shows only the external elements related to a system. This example black box system represents both the 'Focus' and the 'Boundary' it has 'Neighbour', 'Utility', 'Operations' and 'Delivery Project' as 'Environmental elements' which will influence or interact with the System. On many projects there is no differentiation between the 'System' and the 'Delivery Project' but these often have different requirements which are important to understand.

## 11.4   Summary

The Context Pattern defines two Viewpoints that allow aspects of context to be captured. The Context Definition Viewpoint defines the contexts to be considered and the CDVp defines the Focus, Boundary and Environmental elements.

The Context enables a focal aspect related to a system to be considered and is key to ensuring work to be carried out can be understood and bounded.

## 11.5    Related Patterns

When using the Context Pattern other Patterns which may be of use include:

- Description
- Analysis
- Interface
- Test.

## Further reading

Holt, J., Perry, S. and Brownsword, M. *Model-Based Requirements Engineering*. London: IET Publishing; 2011.

*Chapter 12*

# Analysis Pattern

## 12.1 Introduction

The basis for any improvement is analysis. Analysis is the understanding a situation, it may include how it comes about and how it may change over time. When we choose to make an improvement or change we generally have lots of ways we can go about it and we want to take the least risky approach. This pattern provides the core understanding behind any analysis to be carried out. It focuses on cause and effect looking forward and backwards from the change, more specific techniques can be used to further refine the cause or effect within the analysis.

### 12.1.1 Pattern aims

This pattern is intended to be used as an aid to the use of analysis within a Model-Based Systems Engineering (MBSE) application. The main aims of this pattern are shown in the Architectural Framework Context View (AFCV) in Figure 12.1.

The key aim of the Analysis pattern is to allow 'Systems Engineer' to 'Analyse system'. This key aim is constrained by the need to:

- 'Understand context' – the need to understand the point of view from which the system is to be analysed – often this is focused on the scope of a specific discipline such as EMC, RAM, HF or safety.
- 'Manage risk' – the need to manage risk on behalf of the project. This may be focused on cost but could also include technical and other risks.

The key aim 'Analyse system' includes:

- 'Identify issues' – using existing system and input information to identify problems or concerns related to the system, within the relevant context.
- 'Prioritise issues' – using risk techniques to identify those issues which are likely to cause the most problems.
- 'Analyse issues' – using domain specific techniques to understand the issues/ problems, this may include cause, effect, severity, etc.
- 'Suggest mitigations' – provide feedback into the system aim at reducing the issues identified.

Together these aims enable the 'Systems Engineer' to establish in a structured way a set of possible solutions to a specific question.

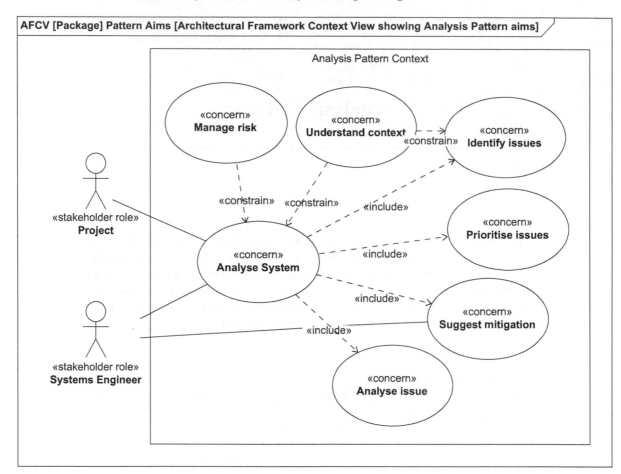

*Figure 12.1   Architectural Framework Context View showing Analysis Pattern aims*

## 12.2   Concepts

The main concepts covered by the Analysis Pattern are shown in the Ontology Definition View (ODV) in Figure 12.2.

Key to this pattern is the concept of an 'Issue'. An 'Issue' provides something to be understood by 'Analysis', 'Issues' are identified from existing 'Model' and can be prioritised by assessment of 'Risk'. 'Analysis' understands the 'Issue' and provides a 'Result' on which a number of 'Recommendation' are based. The intention of the 'Recommendation' is to reduce or resolve the 'Issue'.

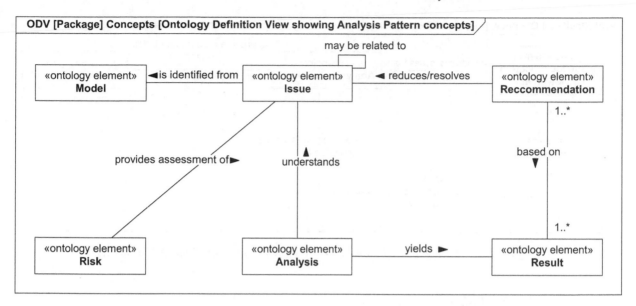

*Figure 12.2   Ontology Definition View showing Analysis Pattern concepts*

## 12.3   Viewpoints

This section describes the Viewpoints that make up the Analysis Pattern. It begins with an overview of the Viewpoints, defines Rules that apply to the pattern and then defines each Viewpoint.

### 12.3.1   Overview

The Analysis Pattern defines a number of Viewpoints as shown in the Viewpoint Relationship View (VRV) in Figure 12.3.

The Analysis Pattern defines four viewpoints for the performance of the Analysis Pattern:

- The 'Issue Identification Viewpoint' is used to collate identified Issues so that they are not lost or forgotten.
- The 'Risk Definition Viewpoint' is used to prioritise the Issues identified.
- The 'Analysis Viewpoint' records the analysis of the Issue, this may be delivered through many different techniques.
- The 'Feedback Viewpoint' records the results and recommendations to be provided based on the Analysis.

Each of these Viewpoints is described in more detail in the following sections. For each Viewpoint an example is also given.

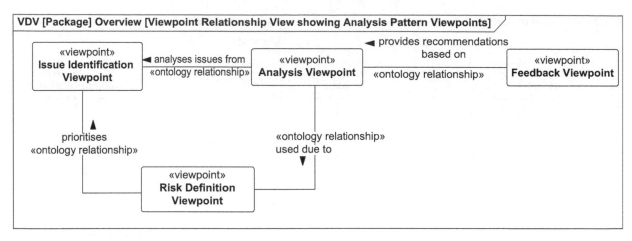

*Figure 12.3   Viewpoint Relationship View showing Analysis Pattern Viewpoints*

### 12.3.2   Rules

Three rules apply to the four Analysis Viewpoints, as shown in the Rules Definition View (RDV) in Figure 12.4.

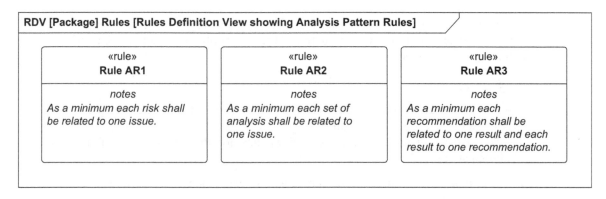

*Figure 12.4   Rules Definition View showing Analysis Pattern Rules*

Note that the three Rules shown in Figure 12.4 are the minimum that are needed. Others could be added if required.

### 12.3.3   Issue Identification Viewpoint (IIVp)

The aims of the Issue Identification Viewpoint are shown in the Viewpoint Context View in Figure 12.5.

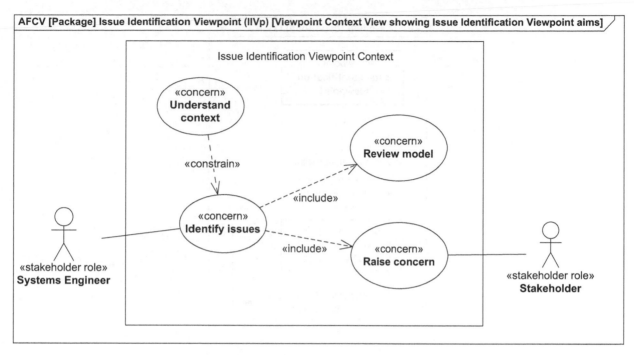

*Figure 12.5   Viewpoint Context View showing Issue Identification Viewpoint aims*

The main aim of the Issue Identification Viewpoint is to 'Identify Issues' which includes:

- 'Review Model' – where concerns will be identified through looking at the model.
- 'Raise Concern' – where stakeholders may identify concerns.

The issues identified will be directed by the 'Understand context' Use Case, this maybe analysis specific.

### 12.3.3.1   Description

The Viewpoint Definition View (VDV) in Figure 12.6 shows the Ontology Elements that appear on an Issue Identification Viewpoint.

The Issue Identification Viewpoint shows the issues and their relationship to each other. Each 'Issue' will have information stored about it including 'Description', 'ID' and 'Raised by'. This information will provide the detail of the issue to be considered. the 'Raised by' ensures the person who originally noted the 'Issue' can be contacted to check exactly what was meant or that any changes which have been made to the issue by others looking at it make sense.

Relationships between 'Issue' could be of many types and the specifics are not defined here so as to leave the relationships to be defined by those defining the 'Issue'.

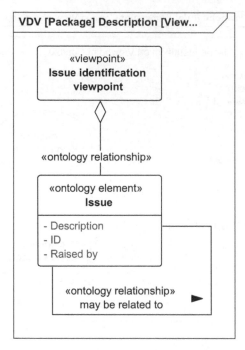

*Figure 12.6   Viewpoint Definition View showing the Ontology Elements that appear on the Issue Identification Viewpoint (IIVp)*

### 12.3.3.2   Example Issue Identification Viewpoint

Example Views that conform to the Issue Identification Viewpoint are shown in Figure 12.7.

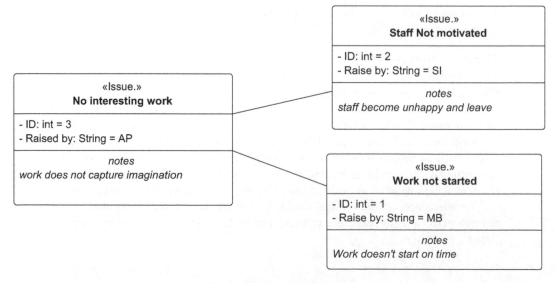

*Figure 12.7   Issue Example*

This example, Figure 12.7 shows three issues which have been raised these all refer to the level and/or type of work to be done in a company, the Description for each is captured in the bottom box for each Issue. In this case the descriptions are short and my need clarification which can be provided by those people referenced by their initials.

The relationships shown here are linked to cause and effect with the perceived cause being on the left and the perceived effect to the right. Of course this could be a circular argument and the real point is that the relationships should help to show the thinking of those who defined the Issues, in this case what may happen if there is 'No Interesting work'.

### 12.3.4   *Risk Definition Viewpoint (RDVp)*

The aims of the Risk Definition Viewpoint are shown in the Viewpoint Context View in Figure 12.8.

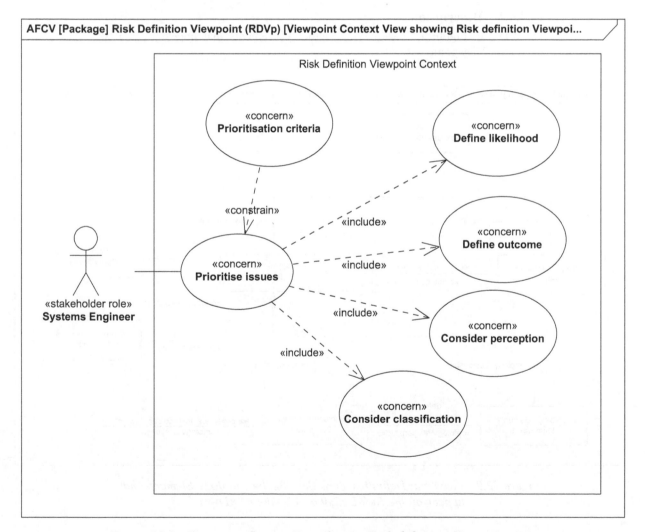

*Figure 12.8   Viewpoint Context View showing Risk definition Viewpoint aims*

The main aim of the Risk Definition Viewpoint is to prioritise issues which includes:

- 'Define outcome' – understanding what the unwanted outcome is based on this issue.
- 'Define likelihood' – understanding what the probability of the outcome occurring is.
- 'Consider perception' – think about what people will say/feel if the outcome happens.
- 'Consider classification' – this can be used if the outcome needs to be grouped or highlighted, such as a safety related outcome.

Issues will be prioritised based on a 'Prioritisation criteria' which may need to be pre-defined.

### 12.3.4.1   Description

The VDV in Figure 12.9 shows the Ontology Elements that appear on a Risk Definition Viewpoint.

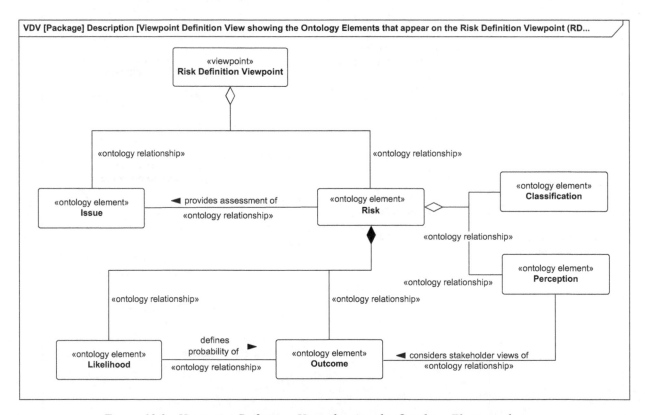

*Figure 12.9    Viewpoint Definition View showing the Ontology Elements that appear on the Risk Definition Viewpoint (RDVp)*

The Risk Definition Viewpoint shows the 'Issue' being prioritised and the 'Risk' associated with it. It also shows the 'Outcome', its 'Likelihood', any stakeholder 'Perception' of the 'Outcome' and any 'Classification' which may be applied.

### 12.3.4.2   Example Risk Definition Viewpoint

An example views that conform to the Risk Definition Viewpoint is shown in Figure 12.10.

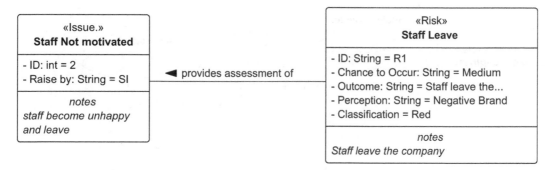

*Figure 12.10   Risk Example*

Figure 12.10 shows a risk defined and associated with one of the Issues from the Issue Identification Viewpoint. In this case only one 'Risk' has been identified and that is the risk of staff leaving, only one risk has been identified as this is the only Outcome of real concern to the organisation others may be on loosing work or making less money. With others a profile of 'Risk' may be shown but in this case the *Red* 'Classification' for the risk means it must be dealt with.

Classifications may be more complex than a Red, Amber, Green as suggested here they could be on safety, security or mission critical terms with another characteristic added to show the priority of the 'Risk'.

### 12.3.5   Analysis Viewpoint (AVp)

The aims of the Analysis Viewpoint are shown in the Viewpoint Context View in Figure 12.11.

The main aim of the Analysis Viewpoint is to 'Analyse system', this includes:

- 'Understand cause' – this ensures any cause related to the issue is understood.
- 'Understand effect' – this ensures any effect of the issue is understood.
- 'Record results' – ensures the analysis is recorded, including the outcome.

The 'Systems Engineer' must ensure all of these aims are met.

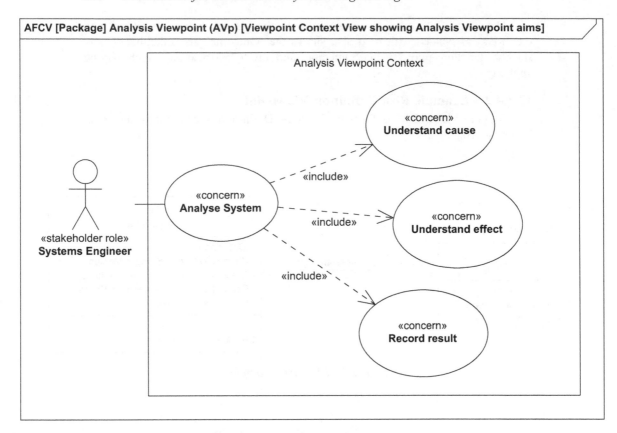

*Figure 12.11   Viewpoint Context View showing Analysis Viewpoint aims*

### 12.3.5.1   Description

The VDV in Figure 12.12 shows the Ontology Elements that appear on an Analysis Viewpoint.

The Analysis view shows the 'Analysis' of the 'Outcome' including any 'Cause Analysis' and 'Effect Analysis'. It also shows the 'Result' of the 'Analysis'.

### 12.3.5.2   Example Analysis Viewpoints

Example Views that conform to the Analysis Viewpoint are shown in Figure 12.13 for Cause Analysis and Figure 12.14 for Effect Analysis.

#### 12.3.5.2.1   Cause Analysis

'Cause Analysis' considers those things which happen leading up to the 'Outcome' in the identified 'Risk'.

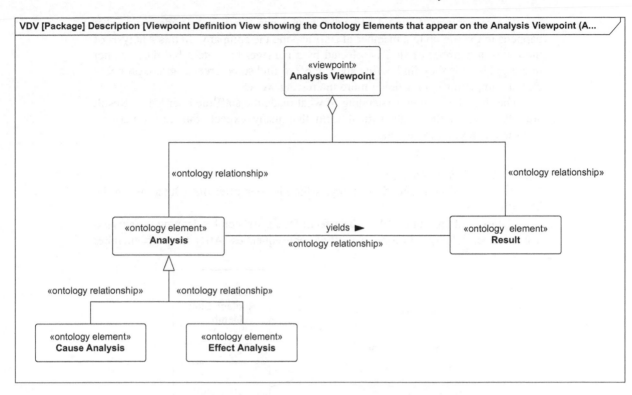

*Figure 12.12    Viewpoint Definition View showing the Ontology Elements that appear on the Analysis Viewpoint (AVp)*

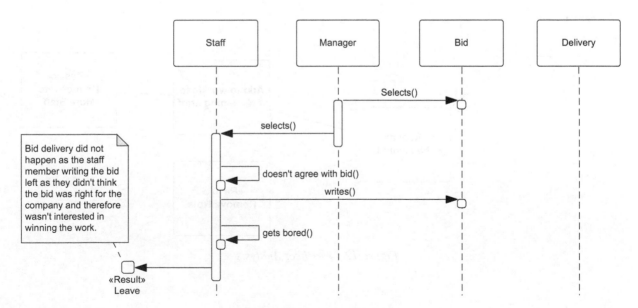

*Figure 12.13    Cause Analysis*

The example in Figure 12.13 uses a sequence diagram to understand what happens in the run up to a member of staff leaving the company. In this Analysis of the Cause a member of staff is selected by a manager to write a bid they are not interested in, this they find boring and run off to find some greener grass, probably with a competitor who is doing more interesting work.

The 'Result' is an understanding of what made the staff member leave. Result are often not in the mathematical form that many expect, this result may be recorded as text on the diagram.

#### 12.3.5.2.2   Effect Analysis

'Effect Analysis' considers those things which happen after the 'Outcome' in the identified 'Risk'.

The example in Figure 12.14 the effect of the staff member leaving is analysed using a tree technique to look at multiple consequences. After the staff member

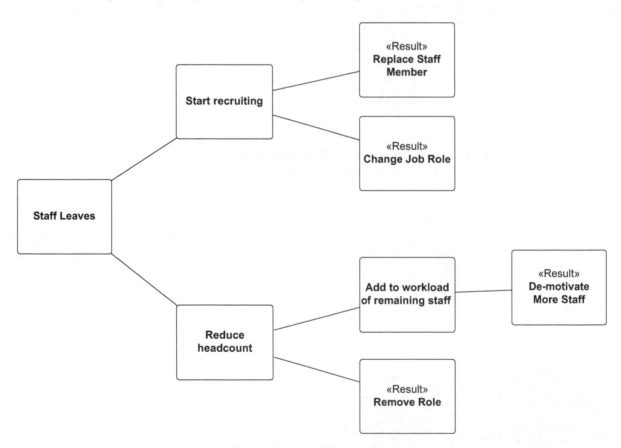

*Figure 12.14   Effect Analysis*

leaves this investigates two options; the option at the top of the diagram looks at what may happen if the company recruits to replace the lost member of staff while the bottom option looks at reducing the headcount noting that some may lead to further de-motivation.

Other consequences could be added to this diagram and further detail considered in terms of probabilities of each branch being taken but this only adds value in specific instances, the value here is to help people recognise what the consequences may be.

The 'Result' in this Analysis is represented by the leaf elements in each branch showing multiple possible 'Result'.

### 12.3.6   Feedback Viewpoint (FVp)

The aims of the Feedback Viewpoint are shown in the Viewpoint Context View in Figure 12.15.

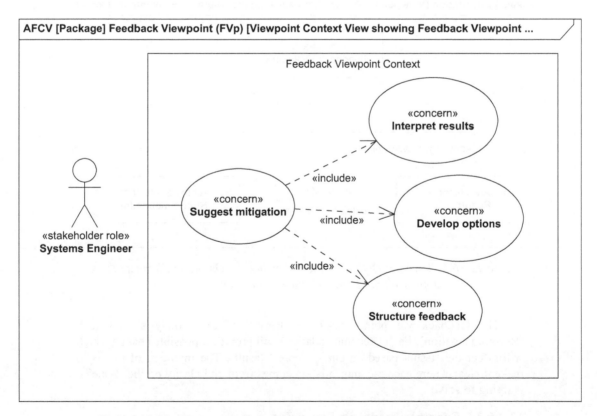

*Figure 12.15   Viewpoint Context View showing Feedback Viewpoint aims*

The main aim of the Feedback Viewpoint is to 'Suggest mitigation' which includes:

- 'Interpret results' – ensuring the results of the analysis are understood and can be used to develop options for mitigations.
- 'Develop options' – designing possible ways to reduce or solve the issue based on the results of the analysis.
- 'Structure feedback' – develop feedback which can be used as a basis for improvements to the overall system.

The 'Systems Engineer' is affected by the mitigations selected, this may mean the role is responsible for delivering the mitigations or implementing them.

### 12.3.6.1    Description

The VDV in Figure 12.16 shows the Ontology Elements that appear on a Feedback Viewpoint.

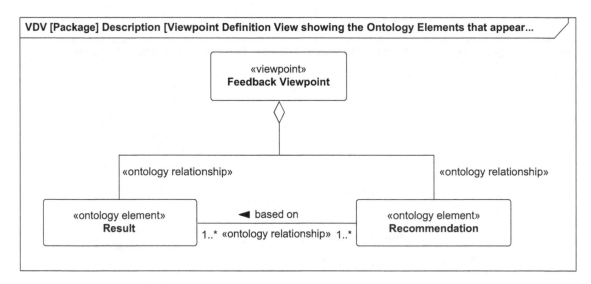

*Figure 12.16    Viewpoint Definition View showing the Ontology Elements that appear on the Feedback Viewpoint (FVp)*

The feedback viewpoint shows the 'Result' of the 'Analysis' and each 'Recommendation'. Each 'Recommendation' will provide a possible change which will affect the 'Issue' based on one or more 'Result'. The intention of this is to make it clear where each recommendation comes from and clarity on the 'Issue' it is trying to solve.

### 12.3.6.2    Example Feedback Viewpoint

An example views that conform to the feedback Viewpoint is shown in Figure 12.17.

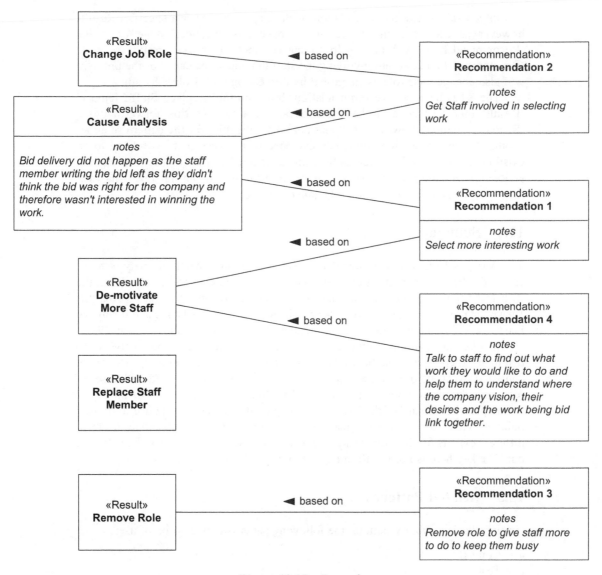

*Figure 12.17    Example*

This example, Figure 12.17 shows four 'Recommendation' based on each 'Result' from the cause and effect analysis. Each 'Result' is taken directly from the Analysis diagrams and collated here, this means that as can be seen the results may provide different levels of detail depending on the type of analysis which has been carried out, this shouldn't be a problem as the results aren't here to be compared, they are here so they can be considered in terms of how they can be used to develop each 'Recommendation' to feedback into the system.

In general each 'Recommendation' is developed to answer a specific 'Result' however as can be seen in the figure, once this has been done often the 'Recommendation' will also help to answer more than one result, such as 'Recommendation 2' where involving the staff to remove the cause of the problem will also change their role as suggested by the 'Change Job Role' 'Result'.

There has been no 'Recommendation' based on the 'Replace Staff Member' 'Result' this may be as it has been decided not to take this forward into a 'Recommendation', however, this does not satisfy the rules for the pattern so either another 'Recommendation' must be developed or this one must be related to an existing recommendation, such as 'Recommendation 3'. Each result must relate to a 'Recommendation' as otherwise we fall into the trap of letting the 'Analysis' make decisions about what should be taken forward.

## 12.4   Summary

The Analysis Pattern defines four Viewpoints that allow aspects of analysis to be captured. The Issue Identification Viewpoint captures the issues to be considered, the Risk Definition Viewpoint defines the risk associated with the issues, the Analysis Viewpoint provides the analysis of risks considered worthy of analysing and finally the Feedback Viewpoint provides recommendations based on the results of the analysis.

It is key to ensure that although experts may be used to carry out analysis and that their expertise can also be used to make decisions that these two aspects are not conflated. Analysis provides an understanding of a system associated with a set of recommendations to improve the system. The 'Recommendation' allows an expert to interpret the 'Result' of the 'Analysis' and provide a focused response from their point of view, this of course may result in different 'Recommendation' from different experts for the same analysis which in themselves will have to be balanced out. The key here is not to fall into the trap of paralysis by Analysis.

## 12.5   Related Patterns

If using the Traceability pattern, the following patterns may also be of use:

- Context
- Test
- Evidence.

## Further reading

Brownsword, M. *'A Formalised Approach to Risk Management'*. Cardiff University, PhD Thesis; 2009.

# Chapter 13
# Model Maturity Pattern

## 13.1 Introduction

Working with models is an ongoing process as the model will evolve as the project progresses. Just because a model exists does not mean that it is fit for purpose or that it has been developed in a logical manner. It is desirable, therefore, to be able to assess the maturity of a model at any point in time.

The concept of maturity is not a new one and is applied in many other areas, such as:

- Processes that may be assessed using a maturity model, such as CMMI (Capability maturity Model Integrated) and ISO 15504 Software Process Assessment.
- People that may be assessed using a competency framework, such as the INCOSE Competency Framework and SFIA.
- Products using Technology Readiness Levels (TRLs) and their derivatives.

The Model Maturity Pattern provides a mechanism for assessing the maturity of any model and, therefore, provides a useful input to any maturity assessment exercise that is model-based. Models can be used to identify, specify, define, evaluate and analyse any artefact, including: software, systems, processes, competencies, requirements, architectures, etc. Therefore, any maturity assessment that may be applied to models, may potentially be applied to any artefact.

### 13.1.1 Pattern aims

This pattern is intended to provide a mechanism to allow the maturity of a model to be assessed. The main aims of this pattern are shown in the Architectural Framework Context View (AFCV) in Figure 13.1.

The key aim of the Model Maturity Pattern is to provide a mechanism to 'Assess the maturity of model'. This need is desirable for any 'System Modeller' and may be applied to any 'Model'. This main Use Case includes the need to 'Define assessment criteria' that may be tailored for any specific application.

There is also a single constraint on this main aim which is that the assessment must 'Apply to any aspect of the model' in three ways:

- 'Apply to ontologies'. The Ontology forms the backbone of any MBSE approach and, therefore, it is very useful to be able to determine how mature that Ontology is.

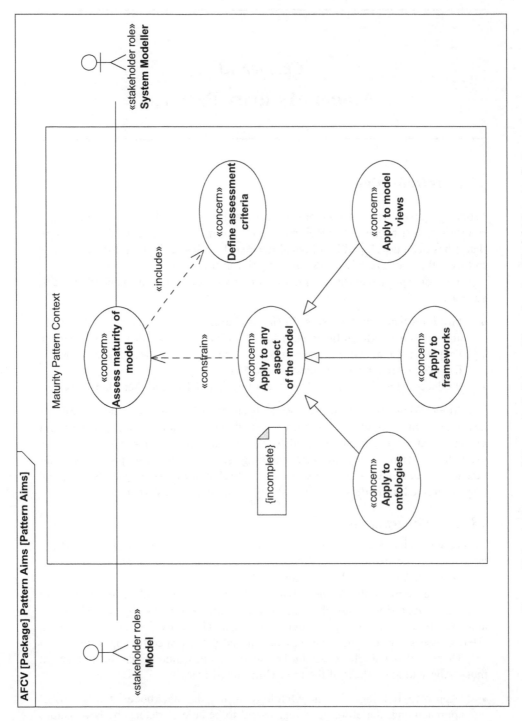

*Figure 13.1   Architectural Framework Context View showing Model Maturity Pattern aims*

- 'Apply to frameworks'. As the Framework contains the Viewpoints that dictate how the Views will be realised, it is important that the maturity of the underlying Framework may be assessed.
- 'Apply to model views'. The actual content of the model, in the form of the Views must also be assessed in terms of its maturity.

This Pattern focuses on defining the assessment, rather than identifying distinct baselines, which is covered by the Epoch Pattern.

## 13.2  Concepts

The main concepts covered by the Model Maturity Pattern are shown in the Ontology Definition View (ODV) in Figure 13.2.

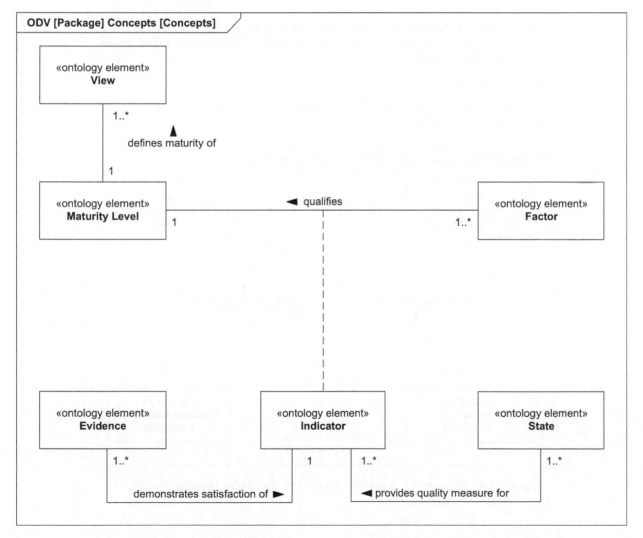

*Figure 13.2   Ontology Definition View showing Model Maturity Pattern concepts*

The diagram here shows the main concepts associated with the Model Maturity Pattern.

The main concept in the Ontology is that of the 'Maturity Level' that defines the maturity of one or more 'View'. Each 'Maturity Level' is qualified by one or more 'Factor' via a set of one or more 'Indicator'. A set of one or more 'Maturity Level' is defined that provides a level of maturity of the model. The set of one or more 'Factor' describes the properties of the model that are being assessed, and the set of one or more 'Indicator' defines what is measured.

Each 'Indicator' is further described by having one or more piece of 'Evidence' associated with it that demonstrates that the Indicator has been satisfied, and one or more 'State' that provides the quality measure for the 'Indicator'.

## 13.3    Viewpoints

This section describes the Viewpoints that make up the Model Maturity Pattern. It begins with an overview of the Viewpoints, defines Rules that apply to the pattern and then defines each Viewpoint.

### 13.3.1    *Overview*

The Model Maturity Pattern defines a number of Viewpoints as shown in the Viewpoint Relationship View (VRV) in Figure 13.3.

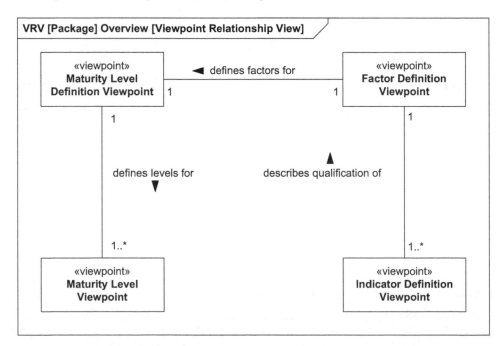

*Figure 13.3    Viewpoint Relationship View showing Model Maturity Pattern Viewpoints*

The Model Maturity Pattern defines four Viewpoints which are:

- The 'Maturity Level Definition Viewpoint' that defines what Maturity Levels exist.
- The 'Factor Definition Viewpoint' that identifies and defines the Factors that are required to determine the Maturity Levels.
- The 'Indicator Definition Viewpoint', that identifies and defines the Indicators and associated types of Evidence and States required to carry out the assessment.
- The 'Maturity Level Viewpoint' that visualises the output of the maturity assessment exercise.

Each of these Viewpoints is described in more detail in the following sections. For each Viewpoint an example is also given.

### 13.3.2   Rules

Five Rules apply to the five Model Maturity Viewpoints, as shown in the Rules Definition View (RDV) in Figure 13.4.

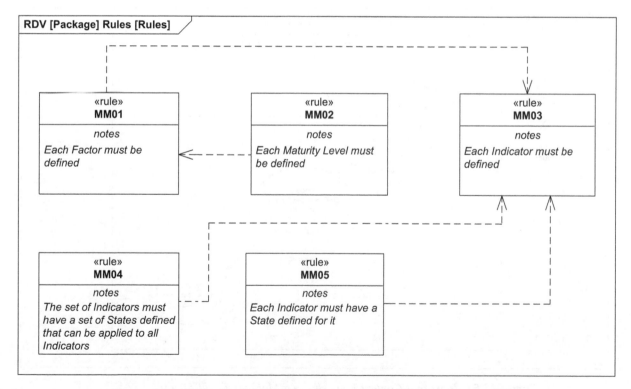

*Figure 13.4   Rules Definition View showing Model Maturity Pattern Rules*

There are five Rules that are applied to the Model Maturity Pattern, which are:

- 'MM01 – Each Factor must be defined', that requires a definition for each of the identified Factors.
- 'MM02 – Each Maturity Level must be defined' that ensures that each Maturity Level has a proper definition.
- 'MM03 – Each Indicator must be defined' that ensures that the Indicators that are used to qualify the Factors is properly defined.
- 'MM04 – The set of Indicators must have a set of States defined that can be applied to all Indicators', that ensures that each Indicator is properly defined in terms of its associated States.
- 'MM05 – Each Indicator must have a State defined for it' that ensures that each Indicator has a State set for it.

Note that the five Rules shown in Figure 13.4 are the minimum that are needed. Others could be added if required.

### 13.3.3   Maturity Level Definition Viewpoint (MLDVp)

The aims of the Maturity Level Definition Viewpoint are shown in the Viewpoint Context View in Figure 13.5.

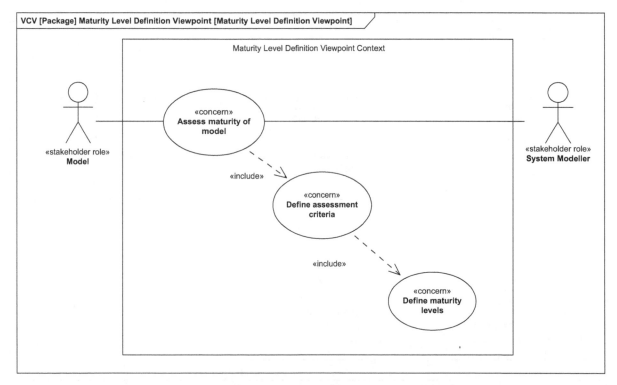

*Figure 13.5    Viewpoint Context View showing Maturity Level Definition Viewpoint aims*

The main aim of the Maturity Level Definition Viewpoint is to contribute to the 'Define assessment criteria' aim which itself contributes to the main aim of the Pattern which is to 'Assess maturity of model'. This is achieved by the 'Define maturity levels' aim.

### 13.3.3.1   Description

The Viewpoint Definition View (VDV) in Figure 13.6 shows the Ontology Elements that appear on a Maturity Level Definition Viewpoint.

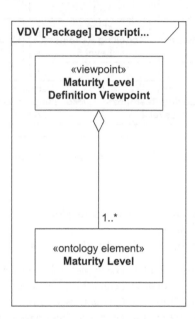

*Figure 13.6    Viewpoint Definition View showing the Ontology Elements that appear on the Maturity Level Definition Viewpoint (MLDVp)*

The Maturity Level Definition Viewpoint is a relatively simple one that consist of a number of Maturity Levels.

### 13.3.3.2   Example Maturity Level Definition Viewpoint

An example View that conforms to the Maturity Level Definition Viewpoint is shown in Figure 13.7.

The Maturity Level Definition View is visualised using a *block diagram* where *blocks* represent the Maturity Levels.

The diagram here shows how a number of Maturity Levels are defined that are being used as part of a Modelling Readiness Level application. The Maturity Levels in this example are called Model Readiness Levels, and are being defined here are used as part of a larger Systems Engineering management activity where a baseline of the Model is taken at various point of the project and then the maturity of the

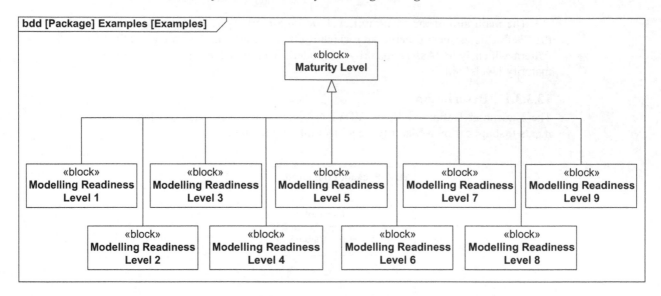

*Figure 13.7    Maturity Level Definition View*

Model is defined. This allows managers to plot the evolution over time of the maturity of the System Model.

There are nine Maturity Levels defined, which are:

- 'Modelling Readiness Level 1: Basic principles observed'
- 'Modelling Readiness Level 2: Technology concept or application formulated'
- 'Modelling Readiness Level 3: Characteristic proof of concept'
- 'Modelling Readiness Level 4: Model defined based on concepts and proof'
- 'Modelling Readiness Level 5: Model validated on relevant test applications'
- 'Modelling Readiness Level 6: Model demonstration in relevant environment'
- 'Modelling Readiness Level 7: Model demonstration in operational environment'
- 'Modelling Readiness Level 8: Model completed and qualified'
- 'Modelling Readiness Level 9: Model proven through successful mission operations.'

This description of each Maturity Level is contained within its *block* on the View which may also be represented as a table. Using a good MBSE approach, this table may be automatically generated from the Model.

This View satisfies Rule MM02.

### 13.3.4    Factor Definition Viewpoint (FDVp)

The aims of the Factor Definition Viewpoint are shown in the Viewpoint Context View in Figures 13.8 and 13.9.

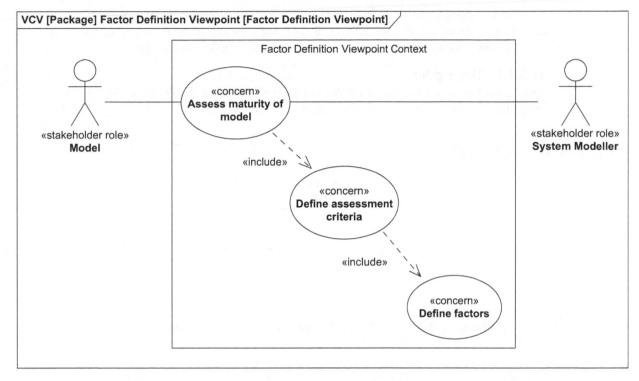

*Figure 13.8   Viewpoint Context View showing Factor Definition Viewpoint aims*

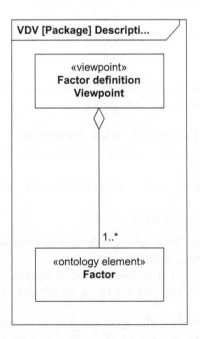

*Figure 13.9   Viewpoint Definition View showing the Ontology Elements that appear on the Factor Definition Viewpoint (FDVp)*

The main aim of the Factor Definition Viewpoint is to contribute to the 'Define assessment criteria' aim which itself contributes to the main aim of the Pattern which is to 'Assess maturity of model'. This is achieved by the 'Define factors' aim.

### 13.3.4.1  Description

The Factor Definition Viewpoint (FDV) in Figure 13.10 shows the Ontology Elements that appear on an Interface Connectivity Viewpoint.

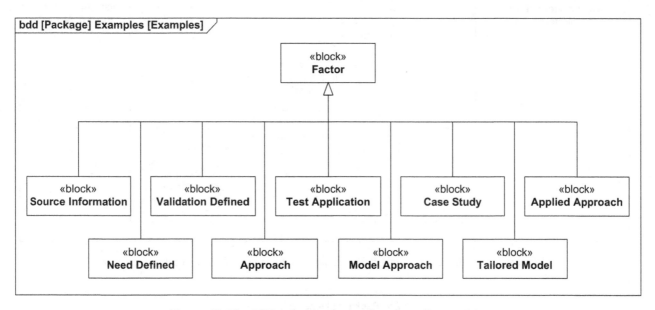

*Figure 13.10   ICV – Service-based Interface Connectivity*

The Factor Definition Viewpoint is another simple View that identifies and defines the Factors required to carry out the assessment exercise.

### 13.3.4.2  Example Factor Definition Viewpoint

Example Views that conform to the Factor Definition Viewpoint are shown in Figure 13.11.

The Factor Definition View is visualised using a *block diagram* where *blocks* represent the Factors.

There are nine Factors defined, which are:

- Source Information: Basic source information is gathered and collated as an input to demonstrating need.
- Need Defined: Concept is defined through definition of needs model.
- Validation Defined: Validation criteria for concept is defined, analysed and reviewed.
- Approach: The model of the approach, based on the concepts and proof is defined and verified.

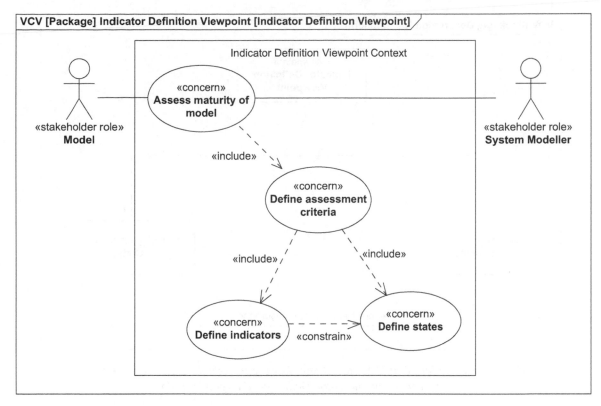

*Figure 13.11    Viewpoint Context View showing Indicator Definition Viewpoint aims*

- Test Application: Partial model of approach is applied to one or more test applications, such as established or predefined test application models.
- Model Demonstration: Model of approach is completed and process is defined. Approach model is applied to one or more test applications, such as established or predefined test application models.
- Case Study: Model is applied to one or more industrial case studies.
- Tailored Model: Model tailored for specific industry and is applied on real industry projects.
- Applied Approach: Model becomes part of industry approach, is measured and controlled by industry quality system.

This description of each Factor is contained within its *block* on the View which may also be represented as a table. Using a good MBSE approach, this table may be automatically generated from the Model.

This View satisfies Rule MM01.

### 13.3.5    Indicator Definition Viewpoint (IDVp)

The aims of the Indicator Definition Viewpoint are shown in the Viewpoint Context View in Figures 13.12 and 13.13.

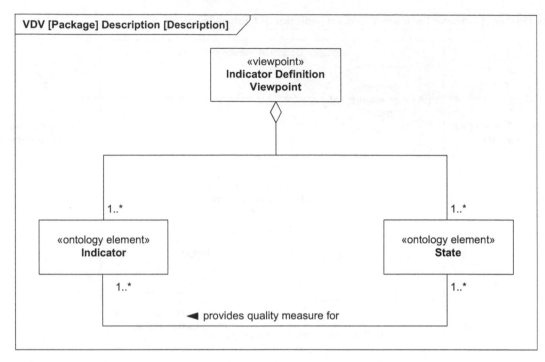

*Figure 13.12    Viewpoint Definition View showing the Ontology Elements that appear on the Indicator Definition Viewpoint (IDVp)*

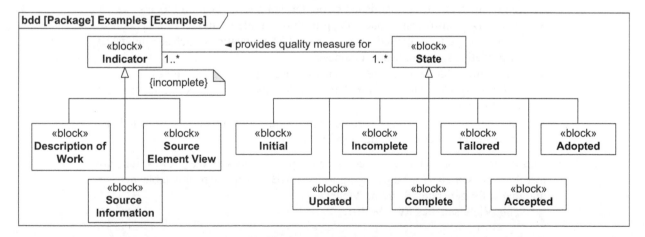

*Figure 13.13    Example of Indicator Definition Viewpoint (IDVp)*

The main aim of the Factor Definition Viewpoint is to contribute to the 'Define assessment criteria' aim which itself contributes to the main aim of the Pattern which is to 'Assess maturity of model'. This is achieved by two aims which are to 'Define indicators' and to 'Define States'.

### 13.3.5.1 Description

The VDV in Figure 13.14 shows the Ontology Elements that appear on an Interface Definition Viewpoint.

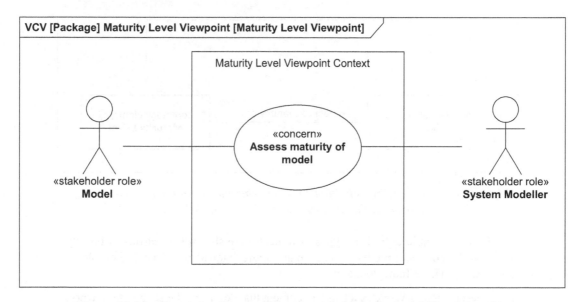

*Figure 13.14 Viewpoint Context View showing Maturity Level Viewpoint aims*

The Indicator Definition Viewpoint is another simple descriptive Viewpoint that defines the set of one or more 'Indicator' and its associated set of one or more 'State'.

Each 'Indicator' shows the assessor what to look for in the Model when performing the assessment, whereas each 'State' provides a qualification of the maturity that can be applied to each 'Indicator'.

### 13.3.5.2 Example Indicator Definition Viewpoint

An example View that conforms to the Indicator Definition Viewpoint is shown in Figure 13.15.

The Indicator Definition Viewpoint is visualised by a *block diagram* where *blocks* are used to represent the Indicators and the States.

The Indicators are used to identify what type of information is required to be used during the assessment. The actual information that is presented by the assessee to demonstrate that the Indicator has been satisfied is the Evidence.

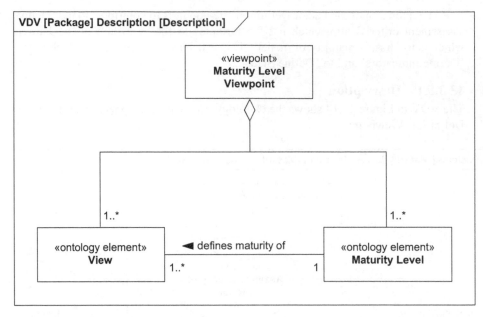

*Figure 13.15    Viewpoint Definition View showing the Ontology Elements that
appear on the Maturity Level Viewpoint (MLVp)*

In this example only three types of Indicator are shown for reasons of brevity whereas the complete model contains many more (indicated by the {incomplete} *constraint*). These Indicators are:

- 'Description of Work', such as a document that describes the scope of the work.
- 'Source Information', such as documents, models, books, emails, etc.
- 'Source Element View', which is, in this case the View taken from the ACRE approach to requirements modelling.

The quality measure for each of these is described by the following States:

- 'Initial' – some artefacts exist but have not been reviewed and are not held under configuration management.
- 'Updated' – the artefacts have been reviewed and, where necessary, updated to reflect the results of the review. Artefacts are held under configuration management.
- 'Incomplete' – the artefacts produced have been reviewed but do not reflect the full set of artefacts for the approach.
- 'Complete' – the artefacts produced have been reviewed and reflect the full set of artefacts for the approach.
- 'Tailored' – the artefacts have been tailored for a specific industry.
- 'Accepted' – the artefacts produced have been reviewed and accepted as fit for purpose.
- 'Adopted' – the artefacts produced have been validated and accepted as fit for purposes and now form part of the industry Quality Management System.

This View satisfies Rules MM03 and MM04.

## 13.3.6 Maturity Level Viewpoint (MLVp)

The aims of the Maturity Level Viewpoint are shown in the Viewpoint Context View in Figure 13.16.

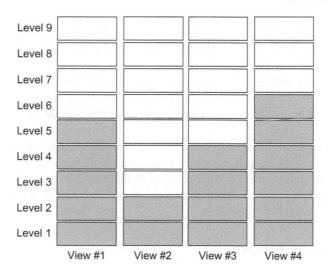

*Figure 13.16 Example of the Maturity Level Viewpoint*

The main aim of the Maturity Level Viewpoint is to 'Assess the maturity of the model'. This aim represents the overall aim of the whole pattern and results in the output of the assessment exercise.

### 13.3.6.1 Description

The VDV in Figure 13.15 shows the Ontology Elements that appear on a Maturity Level Viewpoint.

Figure 13.15 shows the Maturity Level Viewpoint provides the final output of the assessment by showing each relevant 'View' and its associated 'Maturity Level'.

### 13.3.6.2 Example Maturity Level Viewpoint

An example View that conforms to the Maturity Level Viewpoint is shown in Figure 13.16.

Figure 13.16 shows the visualisation of the Maturity Level Viewpoint by using a bar chart. Clearly, this is not a SysML diagram but, providing that the View is consistent with the underlying Pattern definition, then it is still a valid part of the Model.

The x-axis on the bar chart shows the relevant Views that have been assessed, whereas the y-axis shows the nine Maturity Levels that were defined in the Maturity Level Definition View – note that these names are abbreviated for reasons of clarity.

Each cell in the bar chart is either shaded or left blank to indicate which Maturity Level that View has achieved.

This View provides a good indication of the evolution of the maturity of the Model when combined with other instances of the View that have the same set of assessed Views. In order to show the evolution of maturity over time, the Epoch

Pattern should be used to define an Epoch for each baseline of the model under assessment.

## 13.4   Summary

The Model Maturity Pattern provides a mechanism for assessing the maturity of a Model. This assessment may be applied to a whole Model or a specific set of Views that comprise part of the Model.

The Pattern defines four Viewpoints which allow the specification of the Maturity Levels, Factors, Indicators and their associated States.

The resulting Views are relatively simple being represented by *block diagrams* and a bar chart.

This Pattern provides a useful and powerful input to MBSE Project Management activities. When combined with the Epoch Pattern that can be used to define baselines, this Pattern enables the evolution of the Model over time to be plotted. The Maturity Level Views also provide a good input into risk assessment activities for the management of the project.

## 13.5   Related Patterns

If using the Interface Definition Pattern, the following patterns may also be of use:

- Epoch Pattern
- Evidence Pattern.

This does not represent an exhaustive list of the related Patterns and the Patterns may be used in any combination.

## Further reading

Holt, J. and Perry, S. *SysML for Systems Engineering; 2nd Edition: A Model-Based Approach*. London: IET Publishing; 2013.

*Part III*

# Applications of the Patterns

*Chapter 14*

# Requirements Modelling

## 14.1 Introduction

This chapter discusses a Framework for Requirements modelling and shows the enabling Patterns that exist within the Framework. The Approach to Context-based Requirements Engineering (ACRE) Framework has been in use since its early incarnations as far back as 2001 (Holt, 2001) and has, over the years, been extensively published including a complete book describing the approach (Holt *et al.*, 2011; Holt and Perry, 2013). The Framework itself has evolved since its first applications and has been used widely in industry, academia and government for many years.

The existence of the ACRE Framework pre-dates the patterns presented in this book, but was used as one of the sources to identify and define a number of patterns.

## 14.2 Purpose

The purpose of the ACRE Framework is shown in Figure 14.1.

Figure 14.1 shows the overall Context that describes the Use Cases for the ACRE Framework for Requirements modelling.

Figure 14.1 shows that the basic Need for the ACRE Context is to define an approach for requirements engineering ('Support capture of needs'). There are three main constraints for this Use Case, which are:

- The approach that is defined must be model-based ('Must be model-based') which is of interest to the 'MBSE Champion'. This is clearly because this is part of the larger MBSE effort.
- The approach must 'Comply with standards' to ensure that the approach maps onto best practice.
- The approach must 'Identify source of needs' to ensure traceability back to Source Elements.

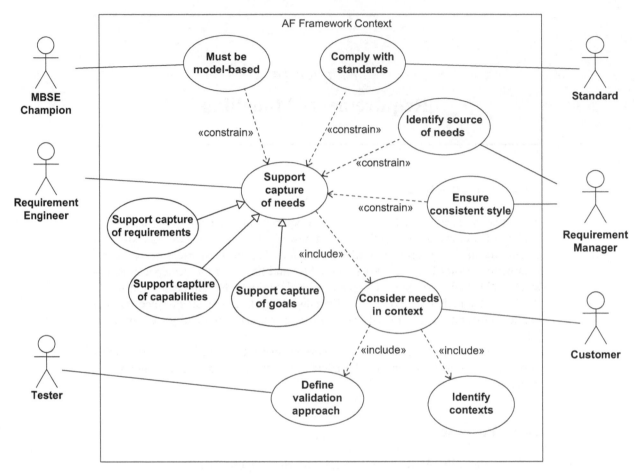

*Figure 14.1    Context for the ACRE Framework*

The main inclusion on this Use Case is to 'Consider needs in context' as the ACRE approach is focussed on context-based modelling, which includes:

- 'Identify contexts', as the approach be able to consider each Need from a specific point of view.
- 'Define validation approach', as all Needs must be validated in order to demonstrate that the System satisfies the original intention.

The whole approach must also be able to cope with Needs that exist at different levels of abstraction ('Support capture of requirements', 'Support capture of capabilities' and 'Support capture of goals').

## 14.3   The ACRE Framework

### 14.3.1   Ontology

This section defines the Ontology for the ACRE Framework, which is shown in Figure 14.2.

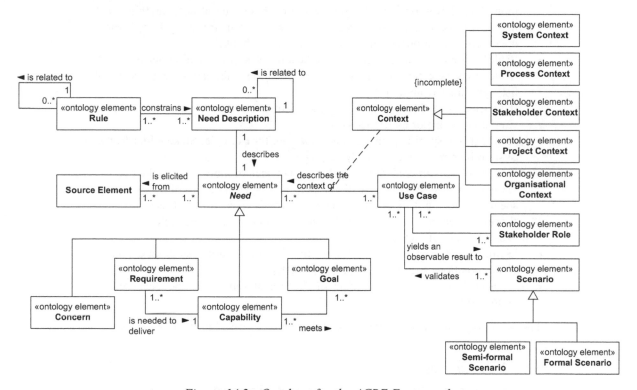

*Figure 14.2   Ontology for the ACRE Framework*

Figure 14.2 shows the MBSE Ontology for the main concepts that are related to Need concepts. These are defined as follows:

- 'Need' – a generic abstract concept that, when put into a 'Context', represents something that is necessary or desirable for the subject of the Context.
- 'Need Description' – a tangible description of an abstract 'Need' that is defined according to a pre-defined set of attributes.
- 'Goal' – a special type of 'Need' whose 'Context' will typically represent one or more 'Organisational Unit' (as an 'Organisational Context'). Each 'Goal' will be met by one or more 'Capability'.
- 'Capability' – a special type of 'Need' whose 'Context' will typically represent one or more 'Project' (as a 'Project Context') or one or more 'Organisational

Unit' (as an 'Organisational Context'). A 'Capability' will meet one or more 'Goal' and will represent the ability of an 'Organisation' or 'Organisational Unit'.

- 'Requirement' – a property of a System that is either needed or wanted by a 'Stakeholder Role' or other Context-defining element. Also, one or more 'Requirement' is needed to deliver each 'Capability'.
- 'Source Element' – the ultimate origin of a 'Need' that is elicited into one or more 'Need Description'. A 'Source Element' can be almost anything that inspires, effects or drives a 'Need', such as: a Standard, a System, Project documentation, a phone call, an email, a letter, a book.
- 'Rule' – is a construct that constrains the attributes of a 'Need Description'. A 'Rule' may take several forms, such as: equations, heuristics, reserved word lists, grammar restrictions.
- 'Use Case' – a 'Need' that is considered in a specific 'Context' and that is validated by one or more 'Scenario'.
- 'Context' – a specific point of view based on, for example, Stakeholder Roles, System hierarchy level, Life Cycle Stage.
- 'Scenario' – an ordered set of interactions between or more 'Stakeholder Role', 'System' or 'System Element' that represents a specific chain of events with a specific outcome. One or more 'Scenario' validates each 'Use Case'.
- 'Formal Scenario' – a 'Scenario' that is mathematically-provable using, for example, formal methods.
- 'Semi-formal Scenario' – a 'Scenario' that is demonstrable using, for example, visual notations such as SysML, tables, text.
- 'Concern' that is a special type of Need that represents the Need of an architecture.

The Viewpoints associated with this Ontology are discussed in the next section.

### 14.3.2 Framework

This section defines the Framework for the ACRE Framework, which is shown in Figure 14.3.

Figure 14.3 shows that there are six main ACRE Viewpoints that are needed to define the Views, which are:

- 'Source Element Viewpoint'. This Viewpoint contains all the Source Elements that are required in order to get the Needs right.
- 'Requirement Description Viewpoint'. This Viewpoint contains structured Need Descriptions. These Need Descriptions are considered individually and will usually have a number of attributes, or features, associated with each one.
- 'Definition Rule Set Viewpoint'. This View contains any Rules that may have to be applied to each Need Description. For example, these may be complexity Rules in the form of equations or more general text-based Rules.
- 'Requirement Context Viewpoint'. This View takes the Needs and gives them meaning by looking at them from a specific point of view by putting them into Context.

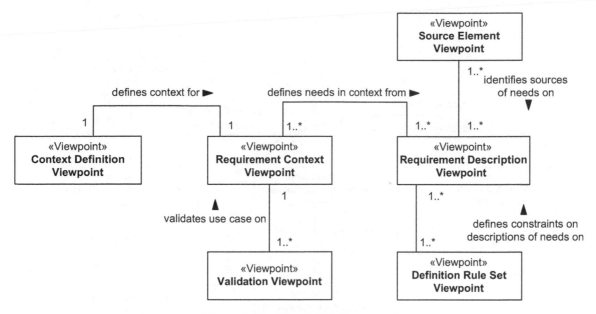

*Figure 14.3    The ACRE Framework*

- 'Context Definition Viewpoint'. This View identifies the points of view that are explored in the Requirement Context Viewpoint.
- 'Validation Viewpoint'. These Views provide the basis for demonstrating that the Needs can be satisfied by defining a number of Scenarios. These Views can describe Semi-formal Scenarios or Formal Scenarios.

These Viewpoints are used as a basis for all of the Views that are created as part of the Model.

### 14.3.3    Patterns

The Enabling Patterns that can be seen in the ACRE Framework are shown in Figure 14.4.

Figure 14.4 shows the original Framework for the "seven views" Framework but this time also shows the Enabling Patterns that have been used.

The Patterns used are as follows:

- Context Pattern that allows the Need Contexts to be defined using the Context Definition Viewpoint and the Requirement Context Viewpoint.
- Testing Pattern that allows the Validation Viewpoints to be realised.
- Description Pattern that allows the Definition Rule Set Viewpoint and Need Description Viewpoints to be defined.

The relationship between the Framework Viewpoints and the Pattern Viewpoints is shown in Table 14.1.

It should also be noted that the Traceability Pattern is used to show all traceability relationships between Ontology Elements.

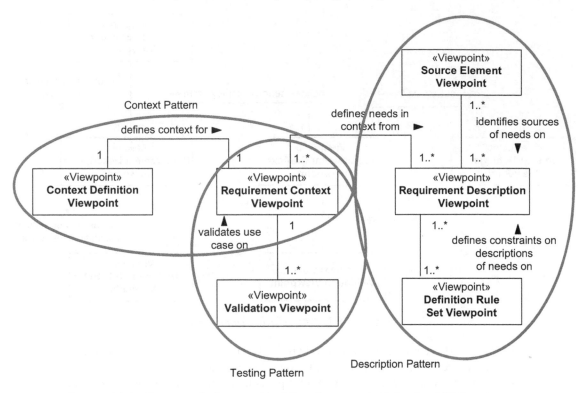

*Figure 14.4    Framework showing Enabling Patterns used for the ACRE Framework*

*Table 14.1    Relationship between Framework and Pattern Views*

| Framework Viewpoint | Related Pattern | Pattern Viewpoint |
|---|---|---|
| Source Element Viewpoint (SEV) | Description Pattern | Element Structure Viewpoint (ESV) |
| Requirement Description Viewpoint (RDV) | Description Pattern | Element Structure Viewpoint (ESV) |
| Definition Rule Set Viewpoint (DRSV) | Description Pattern | Element Structure Viewpoint (ESV) |
| Requirement Context Viewpoint (RCV) | Context Pattern Test Pattern | Context Viewpoint (CV) Test Context Viewpoint (TCV), |
| Validation Viewpoint (VV) | Test Pattern | Element Structure Viewpoint (ESV) |
| Context Definition Viewpoint (CDV) | Context Pattern | Context Viewpoint (CV) |

## 14.4 Discussion

The ACRE Framework uses three main enabling Patterns (the Context, Testing and Description Patterns) plus the Traceability Pattern. The Traceability Pattern is particularly important for the ACRE Framework as traceability is an essential part of any requirements engineering exercise.

The ACRE Framework also shows two Patterns overlapping as the Requirement Context View as a representation of the context appears in both Patterns.

## References

Holt, J. *UML for Systems Engineering – Watching the Wheels*. IET Publishing; 2001.

Holt, J., Perry, S. and Brownsword, M. *Model-Based Requirements Engineering*. IET Publishing; 2011.

Holt, J. and Perry, S. *SysML for Systems Engineering; 2nd Edition: A Model-Based Approach*. London: IET Publishing; 2013.

*Chapter 15*

# Expanded requirements modelling – SoSACRE

## 15.1 Introduction

This chapter discusses the SoSACRE Framework for modelling requirements for Systems of Systems. This Framework is an extension of the ACRE Framework that was described in the previous chapter.

The SoSACRE Framework allows the context of both Systems of Systems and their associated Constituent Systems and allows them to be analysed in terms of consistency between them. The SoSACRE Framework also extends the Validation Views so that they encompass Systems of Systems.

The SoSACRE was developed in 2011 and has been in use ever since on a variety of industry projects.

## 15.2 Purpose

The purpose of the SoSACRE Framework is shown in Figure 15.1.

Figure 15.1 shows that the main Use Case is to 'Provide SoS requirements approach' that must be applicable to different types of Systems of Systems (the constraint 'Apply to different types of SoS') and also across the whole Life Cycle (the constraint 'Apply across life cycle').

There is one main Use Case that helps to realise this, which is to 'Provide SoS requirement engineering processes'. This may at first appear a little odd as there is only a single *include* relationship shown here, but this leaves room for future expansion, for example to define Processes for Requirements management. This has three main inclusions, which are:

- 'Understand context', which applies to both the System of Systems level ('Understand SoS context') and the Constituent System level ('Understand CS context').
- 'Understand relations between CS and SoS', which provides the understanding of the interfaces and interactions between the Constituent Systems and their System of Systems.
- 'Define verification & validation criteria', which ensures that the System of System both works and satisfies its original Needs.

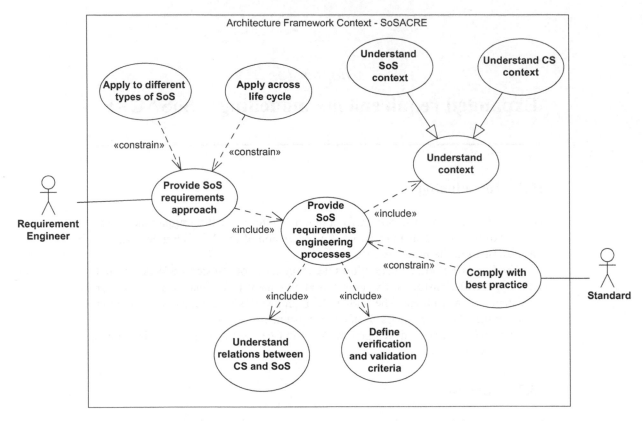

*Figure 15.1    Context for the SoSACRE Framework*

All of this is constrained by the need to meet current best practice ('Comply with best practice').

## 15.3    The SoSACRE Framework

### 15.3.1    Ontology

This section defines the Ontology for the SoSACRE Framework, which is shown in Figure 15.2.

Figure 15.2 shows the MBSE Ontology with a focus on System of System-related elements. These concepts are defined as follows:

- 'System' – set of interacting elements organised to satisfy one or more 'System Context'. Where the 'System' is a 'System of Systems', then its elements will be one or more 'Constituent System', and where the 'System' is a 'Constituent System' then its elements are one or more 'System Element'. A 'System' can interact with one or more other 'System'.
- 'Constituent System' – a special type of 'System' whose elements are one or more 'System Element'.

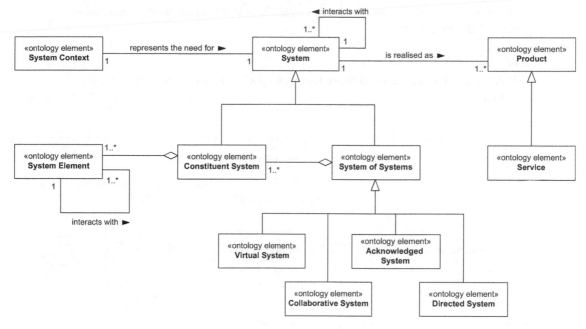

*Figure 15.2  Ontology for the "seven views" Framework*

- 'System of Systems' – a special type of 'System' whose elements are one or more 'Constituent System' and which delivers unique functionality not deliverable by any single 'Constituent System'.
- 'System of Interest' – a special type of 'System' that describes the system being developed, enhanced, maintained or investigated.
- 'Enabling System' – a special type of 'System' that interacts with the 'System of Interest' yet sits outside its boundary.
- 'System Element' – a basic part of a 'Constituent System'.
- 'Product' – something that realises a 'System'. Typical products may include, but are not limited to: software, hardware, processes, data, humans, facilities, etc.
- 'Service' – an intangible 'Product,' that realises a 'System'. A 'Service' is itself is realised by one or more 'Process' *(See also: 'Process')*.
- 'Virtual System' – a special type of 'System of Systems' that lacks central management and resources, and no consensus of purpose.
- 'Collaborative System' – a special type of 'System of Systems' that lacks central management and resources, but has consensus of purpose.
- 'Acknowledged System' – a special type of 'System of Systems' that has designated management and resources, and a consensus of purpose. Each 'Constituent System' retains its own management and operation.
- 'Directed System' – a special type of 'System of Systems' that has designated management and resources, and a consensus of purpose. Each 'Constituent System' retains its own operation but not management.

The next section discusses the changes in the SoSACRE Framework compared to the ACRE Framework.

### 15.3.2   Framework

This section defines the Framework for the SoSACRE Framework, which is shown in Figure 15.3.

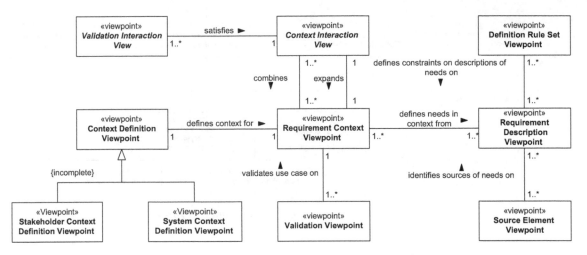

*Figure 15.3    The "seven views" Framework*

Figure 15.3 shows the main Viewpoints that are needed according to the SoSACRE Framework. The basic Viewpoints that have been defined are:

- 'Context Interaction Viewpoint' – intended to provide an overview of the relationships between the Contexts of the various Constituent Systems that make up a System of Systems. Constituent System Context is related to any other Constituent System Context by considering the System of Systems Context and identifying the key relationships.
- 'Validation Interaction Viewpoint' – intended to provide a combined view of the Scenarios for Use Cases that are involved in the System of Systems. Therefore, this Viewpoint combines information from the Validation Viewpoints for various Constituent Systems

The other Views that are based on the standard ACRE Framework and are not described here as they were described in the previous chapter.

### 15.3.3   Patterns

The Enabling Patterns that can be seen in the SosACRE Framework are shown in Figure 15.4.

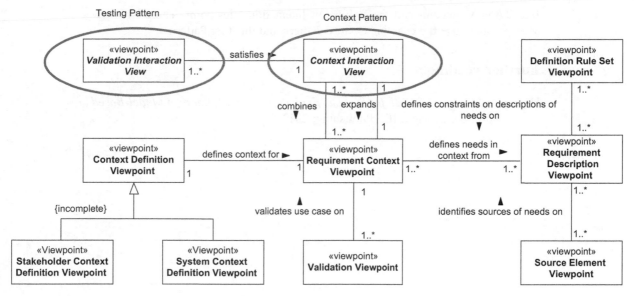

*Figure 15.4 Framework showing Enabling Patterns used for the SoSACRE Framework*

Figure 15.4 shows the original Framework for the SoSACRE Framework but this time also shows the Enabling Patterns that have been used.

The Patterns used are as follows:

- Context Pattern, that allows the Context Interaction View to be defined.
- Test Pattern, that allows the Validation Interaction View to be defined.

The relationship between the Framework Viewpoints and the Pattern Viewpoints is shown in Table 15.1.

*Table 15.1 Relationship between Framework and Pattern Viewpoints*

| Framework Viewpoint | Related Pattern | Pattern Viewpoint |
|---|---|---|
| Context Interaction Viewpoint | Context Pattern | Context Viewpoint (CV) |
| Validation Interaction Viewpoint | Test Pattern | Validation Behaviour Viewpoint (TBV) |

## 15.4 Discussion

The SoSACRE Framework extends the existing ACRE Framework in order to allow the modelling of Needs for Systems of Systems. This is achieved by extending the ACRE Ontology and adding two new Viewpoints: the Context

Interaction Viewpoint and the Validation Interaction Viewpoint. These two new Viewpoints re-use the enabling Context Pattern and the Test Pattern.

## Further reading

Holt, J. and Perry, S. *SysML for Systems Engineering; 2nd Edition: A Model-Based Approach*. London: IET Publishing; 2013.

# Chapter 16

# Process Modelling

## 16.1  Introduction

This chapter discusses the "seven views" Framework for Process modelling and shows the enabling Patterns that exist within the Framework. The "seven views" Framework has been in use since 1999 and was first specified in 2004. The Framework itself has evolved since its first applications and has been used widely in industry, academia and government for many years.

The existence of the "seven views" Framework pre-dates the patterns presented in this book, but was used as one of the source to identify and define a number of patterns.

## 16.2  Purpose

The purpose of the "seven views" Framework is shown in Figure 16.1.

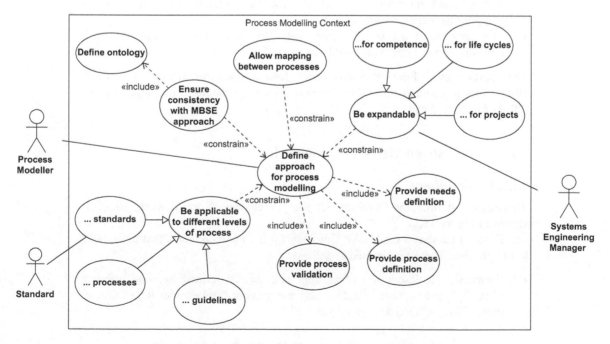

*Figure 16.1  Context for the "seven views" Framework*

Figure 16.1 shows the overall Context that describes the Use Cases for the "seven views" Framework for Process modelling.

- The main use case is concerned with defining an approach to Process modelling ('Define approach for process modelling') which has three main inclusions:
    - There must be a mechanism for allowing the basic Needs of the Process to be defined ('Provide needs definition'). This is a fundamental Need for all aspects of the MBSE approach that is advocated in this book and the area of Process modelling is no exception.
    - Obviously, there must be a mechanism to allow the Process itself to be defined ('Provide process definition'). This will have to meet the basic Needs of the Process that are defined in the previous point.
    - If the Need has been defined and the Process has been defined, then there must be a mechanism in place to allow the Process definition to be validated against the Need ('Provide process validation').

There are four main constraints that are applied to defining the approach, which are:

- The approach must be able to be used with different types of Process that exist at different levels of abstraction ('Be applicable to different levels of process'), such as high-level Standards ('... standards'), medium levels of Process ('... processes') and low-level guidelines ('... guidelines').
- The approach must be expandable for different applications ('Be expandable'). This expandability cover three example areas here: competence ('...for competence'), life cycle modelling ('...for life cycles') and project management ('...for projects').
- The approach must allow for various Processes to mapped together ('Allow mapping between processes').
- The approach must be consistent with the overarching MBSE approach ('Ensure consistency with MBSE approach').

The "seven views" Framework has been extended in several ways to allow modelling of Project Management and Competencies – see chapter 8 of Holt and Perry (2013) for more details.

## 16.3    The "seven views" Framework

### 16.3.1    Ontology

This section defines the Ontology for the "seven views" Framework, which is shown in Figure 16.2.

Figure 16.2 shows the MBSE Ontology for the main concepts that are related to the process. These are defined as follows:

- 'Process' – a description of an approach that is defined by: one or more 'Activity', one or more 'Artefact' and one or more 'Stakeholder Role'. One or more 'Process' also defines a 'Service'.

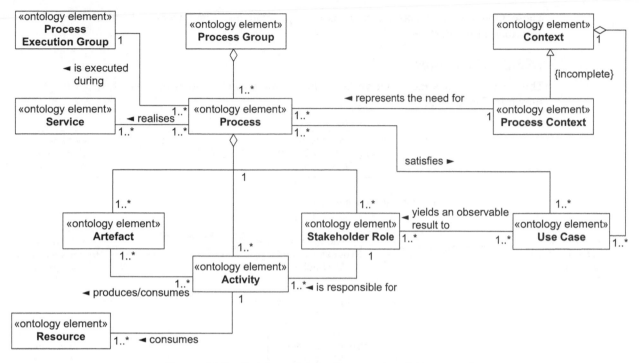

*Figure 16.2    Ontology for the "seven views" Framework*

- 'Artefact' – something that is produced or consumed by an 'Activity' in a 'Process'. Examples of an 'Artefact' include: documentation, software, hardware, systems, etc.
- 'Activity' – a set of actions that need to be performed in order to successfully execute a 'Process'. Each 'Activity' must have a responsible 'Stakeholder Role' associated with it and utilises one or more 'Resource'.
- 'Stakeholder Role' – the role of anything that has an interest in a 'System'. Examples of a 'Stakeholder Role' include the roles of: a 'Person', an 'Organisational Unit', a 'Project', a 'Source Element', an 'Enabling System', etc. Each 'Stakeholder Role' requires its own 'Competency Scope' and will be responsible for one or more 'Activity'.
- 'Resource' – anything that is used or consumed by an 'Activity' within a 'Process'. Examples of a 'Resource' include: money, locations, fuel, raw material, data, people, etc.
- 'Process Execution Group' – a set of one or more 'Process' that are executed for a specific purpose. For example, a 'Process Execution Group' may be defined based on a team, function, etc.
- 'Context' that refers to a specific point of view, in this case this is the 'Process Context' shown by the specialisation.
- 'Service' that refers to any contractual service, whether it is software-based or human-based.

These Ontology Elements are now used to populate the Framework, which is described in the next section.

### 16.3.2   Framework

This section defines the Framework for the "seven views" Framework, which is shown in Figure 16.3.

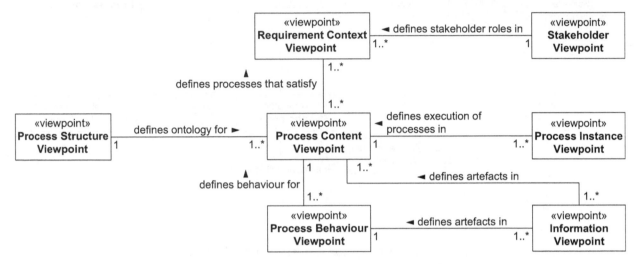

*Figure 16.3   The "seven views" Framework*

Figure 16.3 shows the main Viewpoints that are needed according to the "seven views" approach Framework. The seven basic Viewpoints that have been defined are:

- The 'Requirement Context Viewpoint' (RCV). The 'Requirement Context Viewpoint' defines the Context for the Process, or set of Processes. It will identify a number of Use Cases, based on the Needs for a Process and any relevant Stakeholder Roles that are required.
- The 'Stakeholder Viewpoint' (SV). The 'Stakeholder Viewpoint' identifies the Stakeholder Roles that have an interest in the Processes being defined. It presents Stakeholder Roles in a classification hierarchy and allows additional relationships, such as managerial responsibility, to be added.
- The 'Process Structure Viewpoint' (PSV). The 'Process Structure Viewpoint' specifies concepts and terminology that will be used for the Process modelling in the form of an Ontology. If the Process modelling is taken as part of a larger MBSE exercise, then the Process Structure View will be a subset of the MBSE Ontology.
- The 'Process Content Viewpoint' (PCV). The 'Process Content Viewpoint' identifies the actual Processes, showing the Activities carried out and the Artefacts produced and consumed. The Process Content Viewpoint may be considered as the library of Processes that is available to any Process-related Stakeholder Roles.

- The 'Process Behaviour Viewpoint' (PBV). The 'Process Behaviour Viewpoint' shows how each individual Process behaves in terms of the order of Activities within a Process, the flow of Artefacts through the Process, Stakeholder Role responsibilities and, where relevant, Resource usage.
- The 'Information Viewpoint' (IV). The 'Information Viewpoint' identifies all the Artefacts produced or consumed by Activities within a Process and the inter-relationships between them.
- The 'Process Instance Viewpoint' (PIV). The 'Process Instance Viewpoint' shows instances of Processes in the Process Execution Groups.

These Viewpoints are based on the Enabling Patterns and are described in the following sections.

### 16.3.3 Patterns

The Enabling Patterns that can be seen in the "seven views" Framework are shown in Figure 16.4.

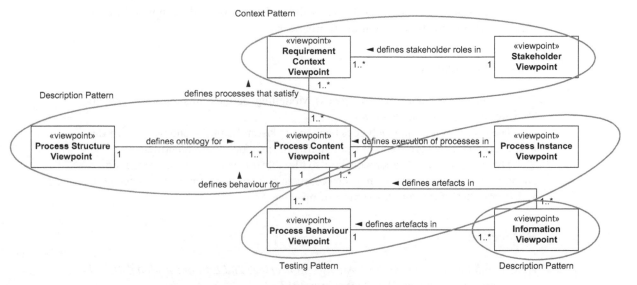

*Figure 16.4 Framework showing Enabling Patterns used for the "seven views" Framework*

Figure 16.4 shows the original Framework for the "seven views" Framework but this time also shows the Enabling Patterns that have been used.

The Patterns used are as follows:

- Context Pattern, that allows the Need for the Process to be expressed.
- Test Pattern, that allows the Process Needs to be validated.
- Description Pattern, that allows the elements in the Information and Ontology Viewpoints to be defined.

The relationship between the Framework Viewpoints and the Pattern Viewpoints is shown in Table 16.1.

*Table 16.1    Relationship between Framework and Pattern Views*

| Framework Viewpoint | Related Pattern | Pattern Viewpoint |
|---|---|---|
| Requirement Context Viewpoint (RCV) | Context Pattern | Context View (CV) |
| Stakeholder Viewpoint (SV) | Context Pattern | Context Definition Viewpoint (CDV) |
| Process Structure Viewpoint (PSV) | Description Pattern | Element Structure Viewpoint (ESV) |
| Process Content Viewpoint (PCV) | Description Pattern | Element Structure Viewpoint (ESV) |
| Process Behaviour Viewpoint (PBV) | Test Pattern | Test Behaviour Viewpoint (TBV) |
| Information Viewpoint (IV) | Description Pattern | Element Structure Viewpoint (ESV) |
| Process Instance Viewpoint (PIV) | Test Pattern | Test Set Behaviour Viewpoint (TSBV) |

It should also be noted that the Traceability Pattern is used to show all traceability relationships between Ontology Elements.

## 16.4    Discussion

The "seven views" approach to Process Modelling defines, unsurprisingly seven Viewpoints and associated Views.

The origin of this approach goes back to the last millennium and was one of the first examples of where some of the Patterns in this book were identified.

The "seven views" approach is also flexible in that it may be extended to encompass other concepts, such as project management and competence – see Holt and Perry, 2013 for a full description of this.

## Reference

Holt, J. and Perry, S. *SysML for Systems Engineering; 2nd Edition: A Model-Based Approach.* London: IET Publishing; 2013.

## Further reading

Holt, J. *A Pragmatic Guide to Process Modelling.* Second edition. Chippenhan, UK: BCS Publishing; 2009.

*Chapter 17*

# Competence Modelling

## 17.1 Introduction

This chapter discusses an approach for competency modelling and assessment, known as the Universal Approach to Competence Assessment and Modelling (UCAM). The UCAM Framework has been in use since 2009 and was first formally published in 2011 (Holt and Perry, 2011) and has been expanded upon since (Holt and Perry, 2013). The Framework itself has evolved since its first applications and has been used widely in industry, academia and government for many years.

The existence of the UCAM Framework pre-dates the patterns presented in this book, but was used as one of the source to identify and define a number of patterns.

## 17.2 Purpose

The purpose of the UCAM Framework is shown in Figure 17.1.

Figure 17.1 shows the overall Context that describes the Use Cases for the UCAM Framework for Competence modelling and assessment.

The main UCAM requirement is to be able to 'Assess competency' which, if it is to be met, includes requirements to 'Understand source framework', 'Populate framework', Set-up assessment' and 'Carry out assessment'.

In order to 'Understand source framework', then there are additional requirements that have to be met to be able to 'Model source framework' and to 'Map source framework to generic framework'.

Competency assessments are not carried out for their own sake (or, at least, *should* not be carried out for their own sake) and so the requirement to 'Assess competency' has a number of variants, showing the need to be able to 'Assess competency for accreditation', 'Assess competency for education', 'Assess competency for recruitment', 'Assess competency for staff appraisal' and 'Assess competency for self-assessment'.

It is no coincidence that the four main areas of UCAM were concerned with understanding any source frameworks to be used (the framework definition elements), populating the frameworks to be used with the information needed to enable them to be used (the framework population elements), determining the reason and scope for an assessment (the assessment set-up elements) and carrying out the

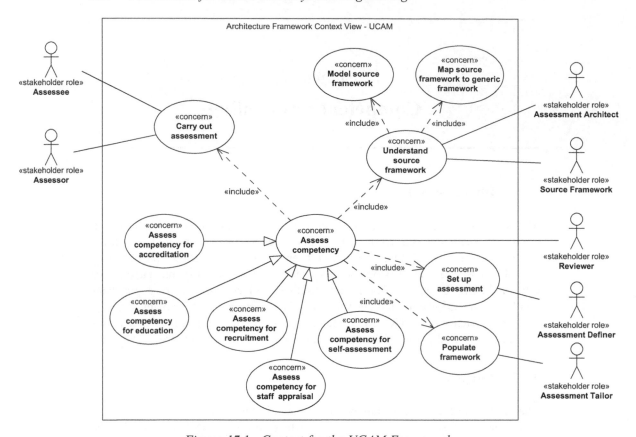

*Figure 17.1    Context for the UCAM Framework*

assessment (the assessment elements). These UCAM processes exist in order to meet the four main UCAM requirements that make up the requirement to 'Assess competency', that is they exist to meet the requirements to 'Understand source framework', 'Populate framework', 'Set-up assessment' and 'Carry out assessment'.

## 17.3    The UCAM Framework

### 17.3.1    Ontology

This section defines the Ontology for the "seven views" Framework, which is shown in Figure 17.2.

Figure 17.2 shows the MBSE Ontology for the main concepts that are related to Competence. These are defined as follows:

- 'Person' – a special type of 'Resource', an individual human, who exhibits 'Competence' that is represented by their 'Competency Profile'. A 'Person' also holds one or more 'Stakeholder Role'.

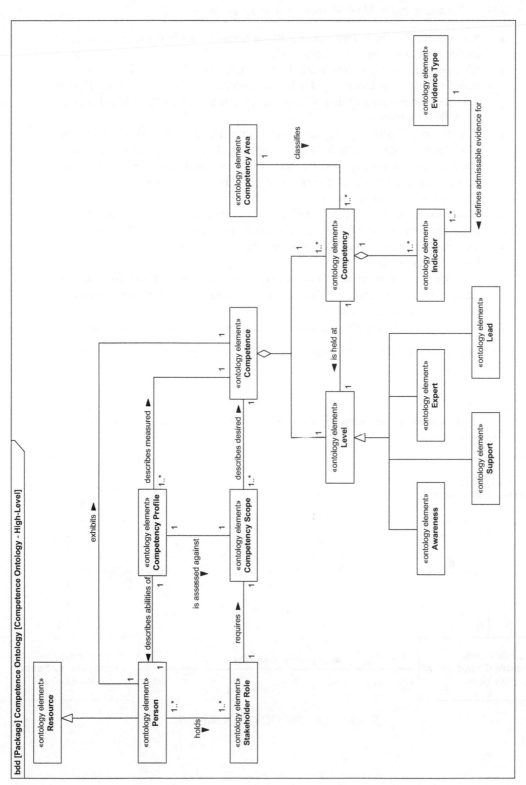

*Figure 17.2   Ontology for the UCAM Framework*

- 'Competence' – the ability exhibited by a 'Person' that is made up of a set of one or more individual 'Competency'.
- 'Competency' – the representation of a single skill that contributes towards making up a 'Competence'. Each 'Competency' is held at a 'Level' that describes the maturity of that 'Competency'. There are four 'Level' defined for the MBSE Ontology.
- 'Competency Profile' – a representation of the actual measured 'Competence' of a 'Person' and that is defined by one or more 'Competency'. An individual's competence will usually be represented by one or more 'Competency Profile'. A 'Competency Profile' is the result of performing a competence assessment against a 'Competence Scope'.
- 'Competency Scope' – representation of the desired 'Competence' required for a specific 'Stakeholder Role' and that is defined by one or more 'Competency'.
- 'Stakeholder Role' – the role of a person, organisation or thing that has an interest in the system.
- 'Indicator' – a feature of a 'Competency' that describes knowledge, skill or attitude required to meet the Competency'. It is the 'Indicator' that is assessed as part of competency assessment.
- 'Evidence Type' – the physical demonstration that an Indicator has been satisfied.

The Competence-related concepts are used to populate the Competence Framework.

## 17.3.2    Framework

This section defines the Framework for the UCAM Framework, which is shown in Figure 17.3.

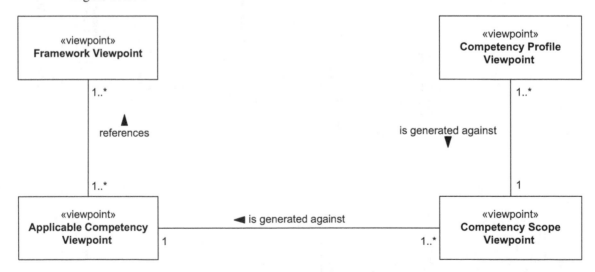

*Figure 17.3    The UCAM Framework*

Figure 17.3 shows the main Viewpoints that are needed according to the UCAM approach Framework. The four basic Viewpoints that have been defined are:

- The main aim of the 'Framework Viewpoint' is to provide an understanding of any source Frameworks that are intended to be used as part of the competency assessment exercise. The Framework Viewpoint is composed of a number of models of source frameworks that can then be mapped to a generic framework.
- The main aim of the 'Applicable Competency Viewpoint' is to define a subset of one or more Competencies that are applicable for a particular Organisation unit. When we create models of Competency Frameworks, it allows us to understand the specific Competencies and the relationships between them. In almost all cases, the set of Competencies in the source Framework will be greater than the Competencies that are relevant for a specific business, therefore, the Applicable Competency Viewpoint contains a pared-down set of Competencies from one or more source Framework.
- The 'Competency Scope Viewpoint' is concerned with identifying and defining a Competency Scope for a specific Stakeholder Role. This is needed as the main input to any competency assessment exercise and provides a definition of the required Competencies and the Levels at which they must be held.
- The 'Competency Profile Viewpoint' is concerned with the actual competencies and the levels that have been measured for a Person. A simple way to think about the two is that the Competency Scope Viewpoint is the main input to a competency assessment exercise, whereas the Competency Profile Viewpoint is the output of such an exercise.

It should be noted that for each of these Viewpoints, an associated set of Views will be generated.

### 17.3.3   Patterns

The Enabling Patterns that can be seen in the UCAM Framework are shown in Figure 17.4.

Figure 17.4 shows the original Framework for the UCAM Framework but this time also shows the Enabling Patterns that have been used.

The Pattern is used are as follows:

- Description Pattern

The relationship between the Framework Viewpoints and the Pattern Viewpoints is shown in Table 17.1.

It should also be noted that the Traceability Pattern is used to show all traceability relationships between Ontology Elements.

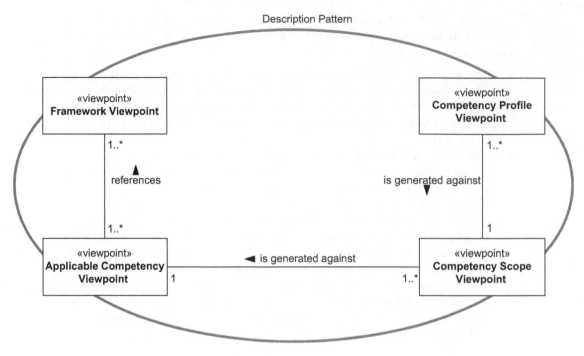

*Figure 17.4    Framework showing Enabling Patterns used for the UCAM*
*Framework*

*Table 17.1    Relationship between Framework and Pattern Views*

| Framework Viewpoint | Related Pattern | Pattern Viewpoint |
|---|---|---|
| Framework Viewpoint (FV) | Description Pattern | Element Description Viewpoint (EDV) |
| Applicable Competency Viewpoint (ACV) | Description Pattern | Element Description Viewpoint (EDV) |
| Competency Scope Viewpoint (CSV) | Description Pattern | Element Description Viewpoint (EDV) |
| Competency Profile Viewpoint (CPV) | Description Pattern | Element Structure Viewpoint (EDV) |

## 17.4    Discussion

The UCAM approach to Competence modelling uses four Viewpoints. This is an unusual application in that all Viewpoints are defined using a single Pattern – the Description Pattern. One of the reasons for this is that the example application here only covers the basis modelling of Competences. When this is performed in real life there will also be an associated set or Processes for carrying out the assessment, which will use the Patterns shown in the Process Modelling Application.

## References

Holt, J. and Perry, S. *SysML for Systems Engineering; 2nd Edition: A Model-Based Approach*. London: IET Publishing; 2013.

Holt, J. and Perry, S. *A Pragmatic Guide to Competency*. Chippenham, UK: BCS Publishing; 2011.

*Chapter 18*

# Life Cycle Modelling

## 18.1  Introduction

This chapter discusses the Life Cycle Modelling Framework that is an expansion of the "seven views" Process Modelling Framework discussed in Chapter 16.

The Life Cycle Framework, like all of the other Frameworks, is flexible in its application and an example of it can be seen in chapter 8 of *SysML for Systems Engineering – A Model-Based Approach*.

Processes have an inherent relationship with Life Cycles, but the nature of this relationship and, indeed, the nature of Life Cycles are often misunderstood. There are many different types of Life Cycle that exist, including, but not limited to:

- Project Life Cycles. Project Life Cycles are perhaps, along with Product Life Cycles, one of the more obvious examples of applications of Life Cycles. We tend to have rigid definitions of the terminal conditions of a Project, such as start and end dates, time scales, budgets, resources and so a Project Life Cycle is one that many people will be able to identify with.
- Product Life Cycles. Again, another quite obvious one is to consider the Life Cycle of a Product. It is relatively simple to visualise the conception, development, production, use and support and disposal of a Product.
- Programme Life Cycles. Most Projects will exist in some sort of higher-level Programme. Each of the Programmes will also have its own Life Cycle and, clearly, this will have some constraint on the Life Cycles of all Projects that are contained within it.
- System procurement Life Cycles. Some Systems may have a procurement Life Cycle that applies to them. From a business point of view, this may be a better way to view a Product, or set of Products, than looking at the Product Life Cycle alone.
- Technology Life Cycles. Any technology will have a Life Cycle. For example, in the world of home computers, the accepted norm for removable storage was magnetic tapes. This was then succeeded by magnetic discs, then optical discs, then solid-state devices and then virtual storage. Each of these technologies has its own Life Cycle.
- Equipment Life Cycles. Each and every piece of equipment will have its own Life Cycle. This may start before the equipment is actually acquired and may end when the equipment has been safely retired. Stages of the equipment life cycle may describe its current condition, such as whether the equipment is in use, working well, degrading and so on.

- Business Life Cycle. The business itself will have a Life Cycle. In some cases, the main driver of the business may be to remain in business for several years and then to sell it on. Stages of the Life Cycle may include expansion or growth, steady states, controlled degradation and so on.

These different types of Life Cycle not only exist, but often interact in complex ways. Also, each of these Life Cycles will control the way that Processes are executed during each Stage in the Life Cycle.

The Life Cycle Modelling Framework is a good example of how the "seven views" Framework has evolved to encompass more than simply Processes. With addition of a few Patterns, it was possible to extend an existing Framework into something with a far wider scope.

## 18.2   Purpose

The purpose of the Life Cycle Modelling Framework is shown in Figure 18.1.

Figure 18.1 shows the overall Context that describes the Use Cases for the Life Cycle Modelling Framework.

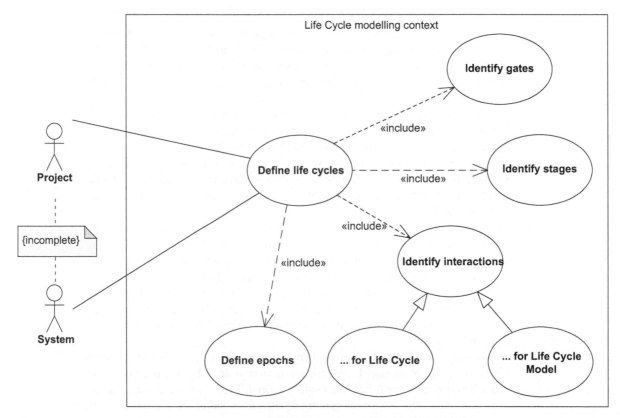

*Figure 18.1   Context for the Life Cycle Modelling Framework*

The main Use Case is concerned with defining Life Cycles ('Define life cycles') which includes four other Use Cases.

The 'Identify stages' and 'Identify gates' Use Case are concerned with identifying the Stage and Gate elements that make up the Life Cycle.

Another key part of this application is to able to model multiple Life Cycles is concerned with identifying interactions ('Identify interactions') in two ways:

- Identifying interactions for Life Cycles ('... for Life Cycle') that are applied to the Life Cycle View.
- Identifying interactions for Life Cycle Models ('... for Life Cycle Model') that are applied to the Life Cycle Model View.

The Framework is also concerned with managing and configuring the Life Cycle, therefore there is a need to define Epochs ('Define Epochs').

The Stakeholder Roles that are shown on the diagram represent some of the types of Life Cycle that may exist. Note that not all of the types of Life Cycle are shown here, as indicated by the {incomplete} constraint.

## 18.3    The Life Cycle Modelling Framework

### 18.3.1    Ontology

This section defines the Ontology for the Life Cycle Modelling Framework, which is shown in Figure 18.2.

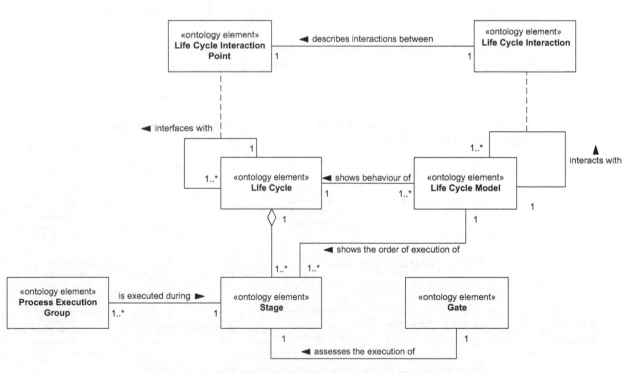

*Figure 18.2    Ontology for the Life Cycle Modelling Framework*

Figure 18.2 shows the MBSE Ontology for the main concepts that are related to the Life Cycle Modelling Framework. These are defined as follows:

- 'Life Cycle' – a set of one or more 'Stage' that can be used to describe the evolution of 'System', 'Project', etc over time.
- 'Life Cycle Model' – the execution of a set of one or more 'Stage' that shows the behaviour of a 'Life Cycle'.
- 'Stage' – a period within a 'Life Cycle' that relates to its realisation through one or more 'Process Execution Group'. The success of a 'Stage' is assessed by a 'Gate'.
- 'Gate' – a mechanism for assessing the success or failure of the execution of a 'Stage'.
- 'Life Cycle Interface' Point – the point in a 'Life Cycle' where one or more 'Life Cycle Interaction' will occur.
- 'Life Cycle Interaction' – the point during a 'Life Cycle Model' at which one or more 'Stage' interact with each other.
- 'Process Execution Group' – an ordered execution of one or more 'Process' that is performed as part of a 'Stage'.
- 'Epoch' that defines a point during the Life Cycle of the Project or System that will form a reference point for Project activities. An Epoch will apply to a specific baseline of the System Model, but not all baselines will be an Epoch. Epochs may be evolved from other Epochs, and may evolve into other Epochs.
- 'Applicable Viewset' that is a subset of the 'System Model' and comprises one or more 'View'. These Views that make up the Epoch will be visualised by one or more 'Diagram'.

The link to the "seven views" Process Modelling Ontology is achieved via the 'Process Execution Group' – see chapter 7 of Holt and Perry (2013).

### 18.3.2  Framework

This section defines the Framework for the Life Cycle Modelling Framework, which is shown in Figure 18.3.

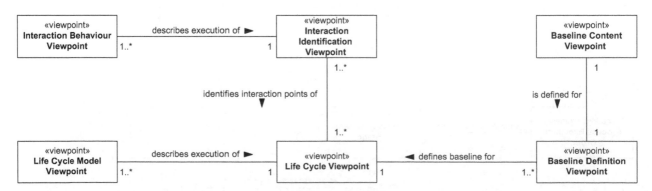

*Figure 18.3   The Life Cycle Modelling Framework*

Figure 18.3 shows that there are six main Viewpoints:

- The 'Life Cycle Viewpoint', that identifies one or more 'Stage' that exists in the 'Life Cycle'.
- The 'Life Cycle Model Viewpoint', that describes how each 'Stage' behaves in relation to one or more other 'Stage'.
- The 'Interaction Identification Viewpoint', that identifies one or more 'Life Cycle Interaction Point' between one or more 'Life Cycle'.
- The 'Interaction Behaviour Viewpoint', that shows the behaviour of each 'Life Cycle Interaction Point' in relation to one or more other 'Life Cycle Interaction Point' as identified in the previous view.
- The 'Baseline Definition Viewpoint' that defines a baseline for a specific Life Cycle.
- The 'Baseline Content Viewpoint' that identifies which Viewpoints from the Model are included in the baseline.

The realisation of these Viewpoints, using the Enabling Patterns, is discussed in the following section.

### 18.3.3 Patterns

The Enabling Patterns that can be seen in the Life Cycle Modelling Framework are shown in Figure 18.4.

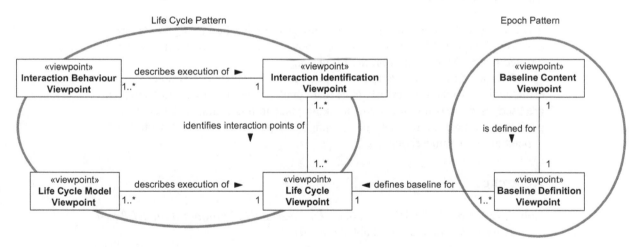

*Figure 18.4    Ontology showing Enabling Patterns used for the Life Cycle Modelling Framework*

Figure 18.4 shows the original Framework for the Life Cycle Modelling Framework but this time also shows the Enabling Patterns that have been used. The Patterns used are as follows:

- Life Cycle Pattern, that allows the Life Cycles, Life Cycle Models and their associated interactions to be defined.
- Epoch Pattern, that allows baselines to be created based on the Life Cycle.

The relationship between the Framework Viewpoints and the Pattern Viewpoints is shown in Table 18.1.

*Table 18.1    Relationship between Framework and Pattern Views*

| Framework View | Related Pattern | Pattern Viewpoint |
| --- | --- | --- |
| Life Cycle Viewpoint (LCV) | Life Cycle Pattern | Life Cycle View (LCV) |
| Life Cycle Model Viewpoint (LCMV) | Life Cycle Pattern | Life Cycle Model View (LCMV) |
| Life Cycle Interaction Viewpoint (LCIV) | Life Cycle Pattern | Life Cycle Interaction View (LCIV) |
| Life Cycle Model Interaction Viewpoint (LCMIV) | Life Cycle Pattern | Life Cycle Model Interaction View (LCMIV) |
| Baseline Definition Viewpoint (BDV) | Epoch Pattern | Epoch Definition Viewpoint (EDV) |
| Baseline Content Viewpoint (BCV) | Epoch Pattern | Applicable Viewset Viewpoint (AVV) |

The Traceability Pattern may be applied to any or all of the relationships shown on Figure 18.4.

## 18.4    Discussion

The Life Cycle Framework has clear and obvious links to the Life Cycle Pattern, as one would expect. The Framework also shows, however, how baselines may be defined for the Model based on the Life Cycle.

The use of the Epoch Pattern here to establish baselines is a simple, yet powerful, addition to the Framework. Note that not all of the Epoch Pattern Viewpoints are used, as there is no measurement taking place in this example. Of course, it is perfectly possible to include the missing Viewpoints from the Epoch Pattern depending on the underlying need.

## Reference

Holt, J. and Perry, S. *SysML for Systems Engineering; 2nd Edition: A Model-Based Approach*. London: IET Publishing; 2013.

*Part IV*

**Using the MBSE Patterns**

*Chapter 19*

# Defining Patterns

## 19.1 Introduction

Like all aspects of model-based systems engineering, successful pattern definition takes practice. However, the pattern definer is not alone. This chapter will discuss the practice of pattern definition, and will provide the novice pattern definer with useful guidance on the practice of pattern definition

After a brief recapitulation of the Framework for Architectural Frameworks (FAF) in the next section, an in-depth discussion of pattern definition follows in Section 19.3. This attempts to show some of the thought processes involved and issues that have to be covered when defining a pattern. This is done using the Description Pattern from Chapter 10 as an example. Section 19.4 then turns to the importance of how Viewpoints are realised, discussing the importance of and giving guidance on examples in pattern definition. The chapter concludes with a summary and references.

## 19.2 The FAF revisited

As discussed in Chapter 3, the approach to pattern definition in this book uses the FAF. This section gives a brief overview of the FAF as a refresher. For each of the FAF Viewpoints, its Viewpoint Definition View (VDV) is given, together with a (very) brief description of its intended purpose and how it may be realised in SysML. It is neither the intention nor the place in this chapter to give a detailed discussion and description of the FAF. For those readers who want such detail see Holt and Perry (2013, chapter 11). The other FAF Views (AFCV, ODV, VRV, RDV and VCVs) are given for reference, without comment, in Appendix B.

The six Viewpoints that make up the FAF are shown in Figure 19.1.

Each of these six Viewpoint is briefly covered in the following sections.

### 19.2.1 AF Context Viewpoint (AFCVp)

The AFCVp is used to address the question of "What is the purpose of the pattern?" The VDV for the FAF AFCVp is given in Figure 19.2. See section 11.2.3.1 of Holt and Perry (2013) for details.

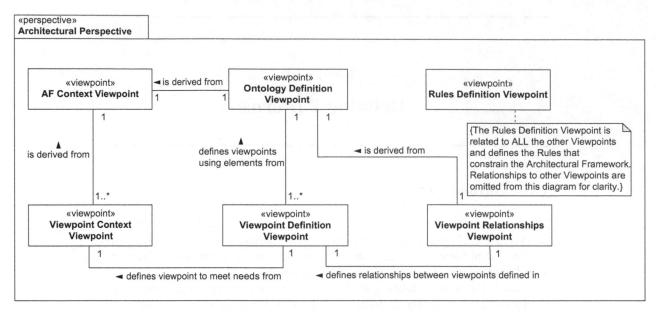

*Figure 19.1    The six FAF Viewpoints*

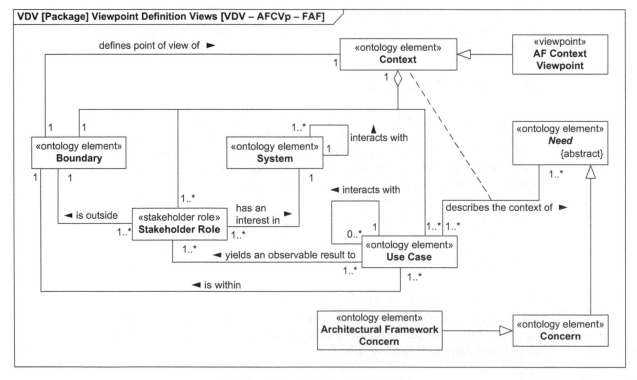

*Figure 19.2    VDV for FAF AFCVp*

An AFCV defines the Context for the pattern, showing the pattern Concerns in context, establishing why the pattern is needed. When using SysML, an AFCV is usually realised as a *use case diagram.*

## 19.2.2 Ontology Definition Viewpoint (ODVp)

The ODVp is used to address the question of "What concepts must the pattern support?" The VDV for the FAF ODVp is given in Figure 19.3. See section 11.2.3.2 of Holt and Perry (2013) for details.

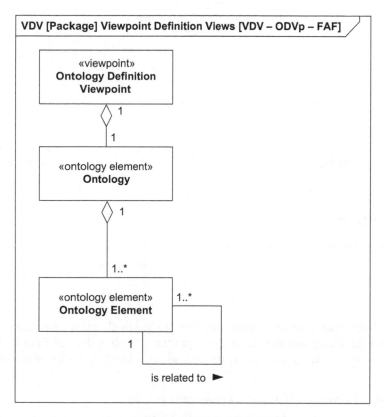

*Figure 19.3 VDV for FAF ODVp*

An ODV defines the Ontology for the pattern. It is derived from the AFCV and defines concepts (Ontology Elements and Ontology Relationships) that can appear on a Viewpoint. When using SysML, an ODV is usually realised as a *block definition diagram.*

### 19.2.3   Viewpoint Relationships Viewpoint (VRVp)

The VRVp is used to address the question of "What different ways of considering the identified concepts are required to fully understand those concepts?" The VDV for the FAF VRVp is given in Figure 19.4. See section 11.2.3.3 of Holt and Perry (2013) for details.

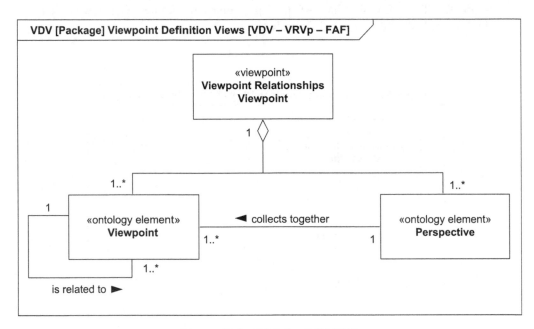

*Figure 19.4    VDV for FAF VRVp*

A VRV identifies the Viewpoints that are required, shows the relationships between the Viewpoints that make up the pattern. It is often derived from the ODV. When using SysML, a VRV is usually realised as a *block definition diagram.*

### 19.2.4   Viewpoint Context Viewpoint (VCVp)

The VCVp is used to address the question of "What is the purpose of each Viewpoint?" The VDV for the FAF VCVp is given in Figure 19.5. See section 11.2.3.4 of Holt and Perry (2013) for details.

A VCV defines the Context for a particular Viewpoint. A VCV represents the Viewpoint Concerns in context for a particular Viewpoint, establishing why the Viewpoint is needed and what it should be used for. It is derived from the AFCV. When using SysML, a VCV is usually realised as a *use case diagram.*

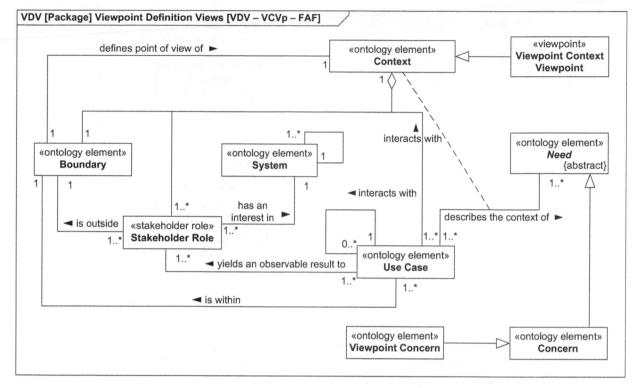

*Figure 19.5 VDV for FAF VCVp*

### 19.2.5 Viewpoint Definition Viewpoint (VDVp)

The VDVp is used to address the question of "What is the definition of each Viewpoint in terms of the identified concepts?" The VDV for the FAF VDVp is given in Figure 19.6. See section 11.2.3.5 of Holt and Perry (2013) for details.

A VDV defines a particular Viewpoint, showing the Ontology Elements and Ontology Relationships (from the ODV) that appear on the Viewpoint. When using SysML, a VDV is usually realised as a *block definition diagram*.

### 19.2.6 Rules Definition Viewpoint (RDVp)

The RDVp is used to address the question of "What rules constrain the use of the pattern?" The VDV for the FAF RDVp is given in Figure 19.7. See section 11.2.3.6 of Holt and Perry (2013) for details.

An RDV defines the various Rules that constrain the pattern. When using SysML, an RDV is usually realised as a *block definition diagram*.

Having given a brief overview of the FAF, we next look at how a pattern is defined in practice.

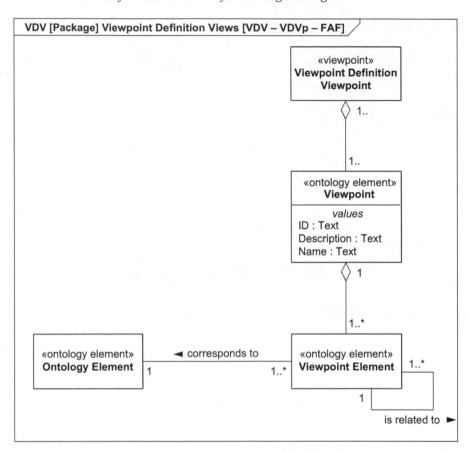

*Figure 19.6   VDV for FAF VDVp*

## 19.3   Defining a Pattern

As discussed in Chapter 3, when defining a pattern there are six essential questions that must be considered. These questions are repeated below:

1. **What is the purpose of the pattern?** It is essential that the purpose and aims of the pattern are clearly understood so that it can be defined in a concise and consistent fashion.
2. **What concepts must the pattern support?** All patterns should be defined around a set of clearly defined concepts that the pattern is defined to address. For example, a pattern that addresses issues of proof might contain the concepts of evidence, claim, argument etc.
3. **What different ways of considering the identified concepts are required to fully understand those concepts?** Often it is necessary to consider the concepts covered by a pattern from a number of different points of view (known in

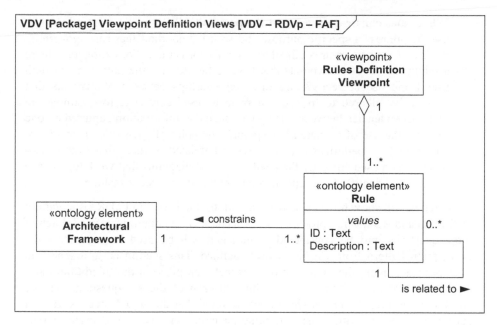

*Figure 19.7  VDV for FAF RDVp*

pattern definition as Viewpoints). A pattern is defined in terms of the View-points that are needed and so such Viewpoints must be clearly identified and any relationships between them identified. For example, a pattern that addresses issues of proof might contain Viewpoints that capture claims made, Viewpoints that show supporting arguments etc.

4.  **What is the purpose of each Viewpoint?** When Viewpoints have been iden-tified as being needed it is essential that the purpose of each Viewpoint is clearly defined. This is needed for two reasons: firstly, so that the pattern definer can define the Viewpoint to address the correct purpose and secondly so that anyone using the pattern understands the purpose of the Viewpoint and can check that they are using the correct Viewpoint (and using the Viewpoint correctly!). Each Viewpoint should have a clear purpose. Good practice also means that each Viewpoint should have, where possible, a single purpose. A common mistake in those with less developed modelling skills is to want to capture all aspects of a problem on a single diagram. This should be avoided.

5.  **What is the definition of each Viewpoint in terms of the identified con-cepts?** Each Viewpoint is intended to represent a particular aspect of the pat-tern and should therefore be defined using only those concepts that have been identified as important for the pattern. Viewpoints can overlap in the infor-mation they present, i.e. concepts can appear on more than one Viewpoint. The Viewpoints must cover all the concepts, i.e. there must be no concept identified as relevant to the pattern that does not appear on at least one Viewpoint.

6. **What rules constrain the use of the pattern?** All patterns are intended to be used to address a specific purpose and so it is important that rules governing the use of the pattern are defined to aid in its correct use. For example, is there a minimum set of Viewpoints that have to be used for the pattern to be both useful and use correctly? Are there relationships between Viewpoints that have to be adhered to for the pattern to be used correctly, for example are there dependencies between Viewpoints such that information captured on one requires the use of a related Viewpoint? The point of such rules is to ensure consistency. A pattern is not just a set of unrelated pictures that exist in isolation; when used correctly the result is a set of diagrams that work together to present the identified concepts in a coherent and consistent fashion.

This section will show how, using these questions, the FAF and PaDRe (the Pattern Definition and Realisation process – see Appendix D), a pattern is defined. Each of these questions will be considered in turn and will be used to show how the Description Pattern from Chapter 10 was defined. This section is, in essence, an attempt to document the thought processes that took place in the mind (and modelling tool!) of the pattern definer. Although each of the six questions will be considered in turn, the reader should bear in mind that the actual process was not carried out in such a linear manner, being far more iterative and involving jumps from question to question and back again. Wherever possible, such jumps will be noted. Also, the style adopted in the following sub-sections will be somewhat more informal than in the rest of the book, with much written in the first person (both singular and plural). Again, this is done in order to give the reader a flavour of the thought processes and working methods of the pattern definer (as to which of the three authors defined this pattern, answers on a postcard!).

### 19.3.1   *What is the purpose of the pattern?*

The first question that has to be addressed is "what is the purpose of the pattern?" When following PaDRe, this is covered by the 'identify context', 'identify source standard' and 'define pattern context' tasks and invocation and exercise of the 'Context Process' within the 'Pattern Definition Process'.

Essentially, the pattern definer is establishing the concerns that the pattern has to address and capturing them as a FAF Architectural Framework Context View (AFCV).

For the Description Pattern, we had long wanted to create a pattern that captured perhaps the most fundamental aspect of model, namely the definition and description of model element. Thinking about how such information is captured in a model (and here, we were thinking about SysML models) suggested the kinds of information that a Description Pattern should allow to be captured, namely:

- The properties and behaviours that a model element may have, For behaviour, we felt that the Description Pattern should be purely structural. This meant that it would allow behaviours to be declared but not defined, i.e. the behaviours that a model element has could be stated but the Description Pattern would not allow the way that they are carried out to be captured.

- The relationships that a model element has to other model elements. Description often relies on context and juxtaposition, so we felt it important that the Description Pattern allow a model element's relationships to be captured. This needed to cover the three common types of relationship, that of decomposition (allowing the way a model element is decomposed into parts) to be captured, taxonomy (allowing "type" relationships to be shown) and general associations between elements.

- Extended, multi-lingual descriptions of model elements. Much modelling, particularly in the software engineering domain, requires textual descriptions of aspects of a system to be localised into a number of languages. We wanted the Description Pattern to support this concept so that a model element could have multiple versions of any descriptive text, each in a different language. We also wanted to allow both brief descriptions and longer descriptions to be created. We also wanted the pattern to capture, as part of these localised descriptions, information on the language used and to do so in a way that was consistent with best practice. This resulted in us deciding that the ISO639 language codes and ISO3166 country codes should be a basis for any such language coding. This in turn led us to the concept of language tags as defined by the Internet Engineering Task Force. For references to these standards, see the 'Further readings' section in Chapter 10.

Having thought about these issues and having identified a number of source standards, we produced the AFCV seen in Section 10.1.1 and reproduced in Figure 19.8.

As mentioned in the introduction to this section, the application of the FAF through PaDRe is, in practice, less linear than both this narrative and PaDRe would suggest and this is the case here. While thinking about the purpose and producing the AFCV we also started thinking about the concepts (discussed in the Section 19.3.2) and some of the rules (discussed in Section 19.3.6); these ideas were captured on initial versions of the ODV and RDV as they occurred to us, rather than waiting for the "correct" stage in the process.

The last point is an important modelling point; the modeller has to be pragmatic and flexible. A modelling process such as PaDRe will work as documented, but is necessarily rather linear and restrictive as it is attempting to define a repeatable set of steps that someone can follow, and that has to work for someone who may have never done the kind of work that the process is defining before. However, once familiar with such a process it becomes, to some extent, internalised and the modeller can execute the process in a more pragmatic and flexible fashion. This does not mean ignoring the process and doing your own thing; all the steps of the process have to be followed correctly otherwise rubbish will result. But, it allows for switching between different parts as necessary according to the practical way that the artefacts being developed by the process evolve. When creating an AFCV it is natural to think about concepts (Ontology Elements in the language of the FAF), so capture them when they occur to you rather than waiting until you get to the invocation of the 'Ontology Definition Process', for example. But, and here is the crucial point, capture them in the way that later process, for example the

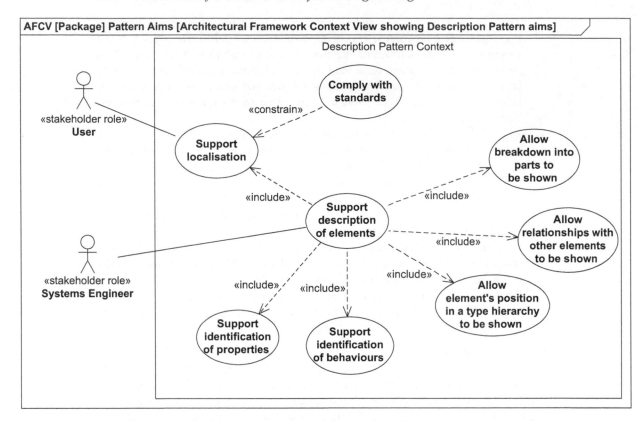

*Figure 19.8   Architectural Framework Context View for the Description Pattern*

'Ontology Definition Process', defines. For us, it meant sketching the start of an ODV and an RDV as the concepts and rules suggested themselves.

The last sentence contains a key word: "sketching". By "sketching" we really do mean applying a drawing instrument to paper or whiteboard. Although the use of a proper modelling tool is essential in defining the patterns fully, the authors' rarely jump straight into their tool of choice. We almost always do all our initial modelling on paper or whiteboards (and sometime on "ePaper" on our tablet devices of choice) before firing up our SysML tool.

So, in summary, whenever you think you might have a new pattern, ask yourself exactly what the purpose of the pattern is. Write down your thoughts on the purpose and think about what you have written some more. You are aiming to clearly articulate the purpose of the pattern. Document the purpose using an AFCV. At this point, stop and think some more. Are there too many use cases? If there are, then perhaps you have multiple patterns or even the beginning of a framework. You want each pattern to have a clearly defined and well-bounded purpose. Also, do not be afraid to capture other information as it occurs to you. It is quite common, when producing an AFCV, to start thinking about concepts that the pattern will need to support and also possible rules on how it will be used. Start to capture these thoughts on the appropriate View (for example, on an ODV or RDV).

When you are happy that you have captured the purpose of the pattern, it is time to address the second question.

### 19.3.2   What concepts must the pattern support?

The second question that has to be addressed is "what concepts must the pattern support?" When following PaDRe, this is covered by the 'define pattern ontology' task and invocation and exercise of the 'Ontology Definition Process' within the 'Pattern Definition Process'.

Essentially, the pattern definer is establishing the concepts that the pattern is defined to address and capturing them as a FAF Ontology Definition View (ODV).

When defining the Description Pattern, three concepts immediately suggested themselves based on the purpose of the pattern established in the AFCV:

- The concept of an "element" that forms the main modelling concept that the pattern addresses. All elements need to be able to be named and described, and so must have "name" and "description" properties as standard.
- The concept of a "property" so that an element can be defined in terms of its defining properties.
- The concept of a "behaviour" so that an element can be defined in terms of its provided behaviours.

These three concepts cover the 'Support description of elements', 'Support identification of properties' and 'Support identification of behaviours' Use Cases from the AFCV that was produced for the Description Pattern. Indeed, the initial version of the ODV, containing Ontology Elements for these concepts, was produced while creating the AFCV.

Looking at the AFCV also immediately identifies three different relationships that must be able to be created between elements, namely relationships that show:

- Elements being broken down in to sub-elements.
- Elements related to other elements through taxonomy ("type of") relationships.
- General relationships not covered by the other two kinds.

These three relationships cover the 'Allow breakdown into parts', 'All element's position in a type hierarchy to be shown' and 'Allow relationships with other elements to be shown' Uses Cases from the AFCV that was produced for the Description Pattern. Again, Ontology Relationships were added to the initial ODV while creating the AFCV.

These concepts and their relationships were captured on an initial version of an ODV. However, this still left two Use Cases from the AFCV that were not covered by the Ontology Elements and Ontology Relationships on the ODV, namely:

- 'Support localisation'
- 'Comply with standards'.

Thinking about these Use Cases led to the introduction of another Ontology Element to the ODV, that of an "element description". The idea is that the "element description" is used to capture localised descriptions of an "element". Again, by looking at the Use Case on the AFCV, the "element description" had to have properties to capture brief and full descriptions. Also, given that we wanted to comply with standards in terms of identifying the languages used, the "element description" also needed a property to capture a language code. We also, at this point, decided that an "element", when described through an "element description", have a "presentation name". This would allow the name of an "element" to be localised (i.e. translated) with the translation captured in the "presentation name".

Putting all these concepts together, and relating "element description" to "element" gives the complete ODV, as found in Section 10.2 and reproduced in Figure 19.9.

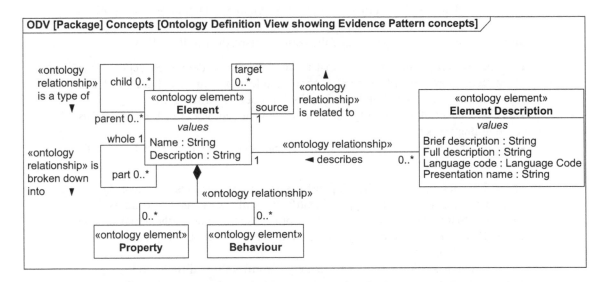

*Figure 19.9   Ontology Definition View for the Description Pattern*

In summary, take the AFCV and, thinking about the Use Cases on the AFCV, create an ODV containing Ontology Elements and Ontology Relationships that capture the concepts that are needed to address the Use Cases. Be prepared to begin work on the ODV while working on the AFCV; as discussed previously, this is perfectly fine and is the way that the authors work. Also, remember that the ODV forms the heart of the pattern; the Viewpoints that the pattern defines can use only (and all) those concepts that appear on the ODV. This might mean that you will need to revisit the ODV later in the process. Perhaps, when defining a Viewpoint, new concepts will be discovered that must be added to the ODV. Or, conversely, when all Viewpoints have been defined you find that there are concepts that have not been included in any Viewpoint. If you haven't missed a Viewpoint, then these concepts should be removed from the ODV.

So, you have defined what the pattern is for and identified the concepts that the pattern needs to address. It is now time to address the third question.

### 19.3.3 *What different ways of considering the identified concepts are required to fully understand those concepts?*

The third question that has to be addressed is "what different ways of considering the identified concepts are required to fully understand those concepts?" When following PaDRe, this is covered by the 'identify viewpoints' task within the 'Pattern Definition Process'. Viewpoint identification does not have a separate Process in PaDRe.

Essentially, the pattern definer is identifying a set of candidate Viewpoints that will make up the pattern and is capturing them as a FAF Viewpoint Relationship View (VRV).

The best way to approach this is to look at copies of the AFCV and ODV. Starting with the AFCV, look for sensible groupings of Use Cases. Each such grouping is a candidate Viewpoint. Circle these groupings. When we do this we use printed copies and mark the AFCV, circling the Use Cases that we think are related. For the Description Pattern, our initial groupings were those shown on Figure 19.10.

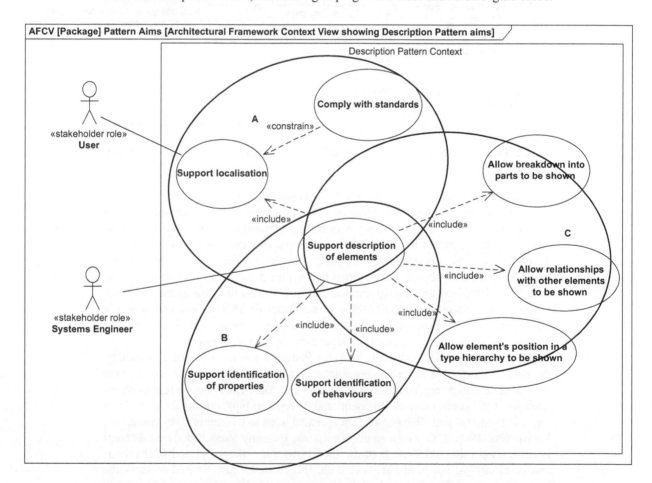

*Figure 19.10   AFCV showing possible Viewpoints through grouping of Use Cases*

As shown in Figure 19.10, we initially identified three candidate Viewpoints, labelled 'A' to 'C' in the diagram. Next, look at the ODV and in a similar fashion circle the Ontology Elements and Ontology Relationships that support each of the possible Viewpoints. Again, paper is best. A marked-up ODV based on the AFCV above is shown in Figure 19.11.

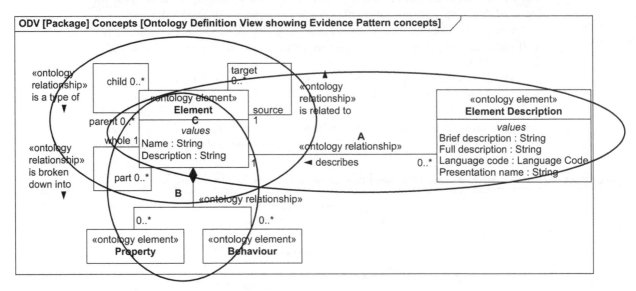

*Figure 19.11    ODV showing Ontology Elements on each of the possible Viewpoints*

So now you have a set of candidate Viewpoints. Look for significant overlaps in the Ontology Elements and Ontology Relationships that will appear on each Viewpoint. There is nothing wrong with having Ontology Elements appearing on multiple Viewpoints; in fact, this is quite common. However, ask yourself whether a significant overlap is giving you anything. Do you need the separate Viewpoints or could they be combined. Don't forget that a Viewpoint defines what can appear on a View based on that Viewpoint. Careful definition will allow some Ontology Elements and Ontology Relationships to be optional. This means that a single Viewpoint can be realised by multiple Views, each with a slightly different emphasis, thus avoiding the need for separately defined Viewpoints.

When we were defining the Description Pattern we were immediately struck by the significant overlap between Viewpoints 'B' and 'C'. Viewpoint 'B' could show an 'Element' and its breakdown into 'Property' and 'Behaviour' but no relationships. Viewpoint 'C' could show an 'Element' and its relationships, but not its breakdown into its 'Property' and 'Behaviour'. This seemed to us to be limiting. By combining Viewpoints 'B' and 'C' into a single Viewpoint, then any View based on this combined Viewpoint could show both the breakdown into 'Property' and 'Behaviour' and relationships, but need not show both. If a modeller just wanted to show the breakdown into 'Property' and 'Behaviour', then the Viewpoint would allow it.

If just relationships were to be shown, then that could be done too. And if both were wanted, then that would also be possible using the same Viewpoint.

We revised our mark-up of the AFCV as shown in Figure 19.12.

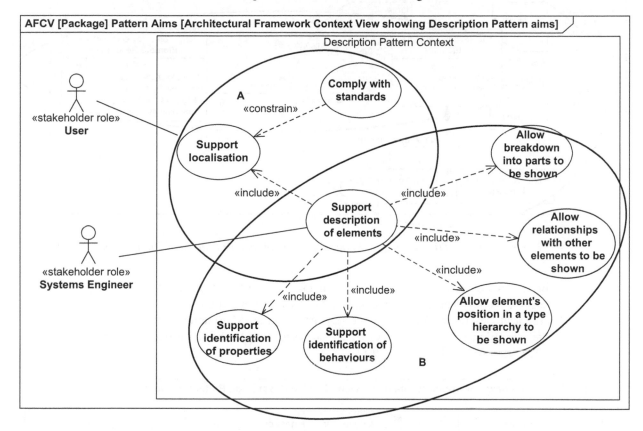

*Figure 19.12   AFCV showing final Viewpoints through grouping of Use Cases*

And similarly, we revised our mark-up of the ODV as shown in Figure 19.13.

A note of caution should be made here. What you consider "significant overlap" is, of course, subjective. Why not have a single Viewpoint that covers all the concepts, for example? There is no easy answer to this. Try to be pragmatic, think whether the groupings of Use Cases feel "natural" and always remember that patterns are there to help and so should be easy to understand and use. A single Viewpoint which can be used with different emphasis in a couple of ways seems reasonable. One that can be used in 10 different ways would suggest that you would be better having four or five different Viewpoints.

The final step is naming the Viewpoints and creating the VRV. Try to keep the names short. Views based on Viewpoint will probably be known by an acronym based on their name (at least, we always do that. For example, we always talk about "ODVs" and rarely use the full name). For this reason, Viewpoints with two, three

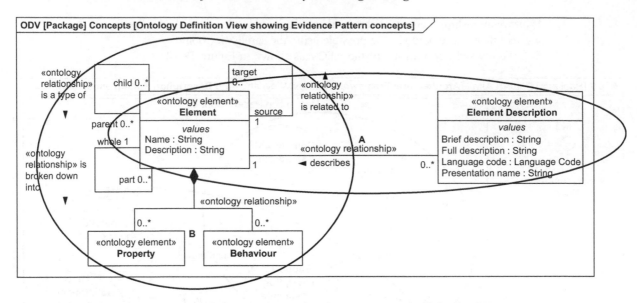

*Figure 19.13   ODV showing Ontology Elements on each of the final Viewpoints*

or four word names work best. The groupings of Use Cases on the AFCV may suggest names. The two Viewpoints in the Description Pattern were named the 'Element Description Viewpoint' (corresponding to the 'A' grouping on the AFCV) and the 'Element Structure Viewpoint' (corresponding to the 'B' grouping). The VRV is reproduced in Figure 19.14.

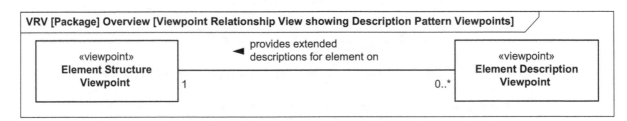

*Figure 19.14   Viewpoint Relationship View for the Description Pattern*

So, in summary, use the AFCV to identify logical groupings of Use Cases. Each such grouping is a candidate Viewpoint. Mark up the ODV to show the Ontology Elements and Ontology Relationships corresponding to these groupings. Look at the marked up ODV and think about overlap. Can a single Viewpoint achieve the same as multiple Viewpoints? Beware, however, too much combining of Viewpoints. When you are happy with the candidate Viewpoints, revise the AFCV and ODV groupings as appropriate. Name your Viewpoints and produce a VRV.

At this point, keep your marked up AFCV and ODV as these annotated diagrams will prove invaluable when answering the fourth and fifth questions. With the Viewpoints identified it is time to start defining them. First, we will think about the purpose of each Viewpoint by asking the fourth question.

### 19.3.4    What is the purpose of each Viewpoint?

The fourth question that has to be addressed is "what is the purpose of each Viewpoint?" When following PaDRe, this is covered by the 'Viewpoint Definition Process' which invokes the 'Context Process'.

Essentially, the pattern definer is identifying the concerns that each Viewpoint has to address and is capturing them as a FAF Viewpoint Context View (VCV).

Each Viewpoint will address a subset of the concerns found on the AFCV. What concerns does each Viewpoint address? Well, exactly those Use Cases grouped on the AFCV that was marked up in order to help identify the Viewpoints. This is why keeping the marked up AFCV is so useful. Essentially in identifying the Viewpoints you have already identified the concerns. So, take the marked up AFCV and create VCVs for each of the Viewpoints identified, showing only the relevant Use Cases for each Viewpoint.

When doing this in a SysML tool, the authors copy the AFCV as many times as there are identified Viewpoints, rename the copies to be VCVs and delete the Use Cases that are not relevant to a particular Viewpoint. The VCV for the Element Structure Viewpoint is reproduced below in Figure 19.15.

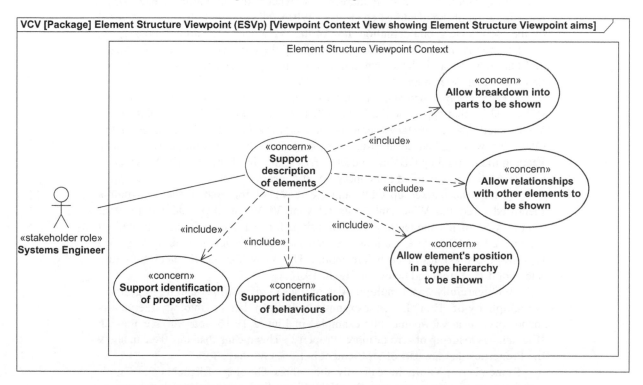

*Figure 19.15    Viewpoint Context View for the Element Structure Viewpoint*

Sometimes, simple copying is not enough. For more complicated Viewpoints it might be necessary to further analyse some of the Use Cases on a VCV, breaking them down further through the use of inclusion, extension or constraint relationships to more detailed Use Cases. If this is done, then these additional Use Cases will appear on the VCV only; they are not typically added back to the AFCV. Of course, if in creating the VCVs you realise that there are other concerns that should have been handled by the pattern but which has been missed, it is necessary to revisit the AFCV and go through questions 1–3 again.

In summary, take the marked-up AFCV used to identify Viewpoints and create a VCV for each Viewpoint. As a starting point, each VCV will contain the Use Cases that were grouped for that Viewpoint on the marked-up AFCV. If necessary, expand on the Use Cases on each VCV. If missing concerns are spotted, add them to the AFCV and start again by revisiting the AFCV and re-asking question 1.

Although the PaDRe 'Viewpoint Definition Process' says to create a VCV and then a VDV for each Viewpoint in turn, you may find it easier to do all the VCVs first. Whichever works best for you is fine.

### 19.3.5    What is the definition of each Viewpoint in terms of the identified concepts?

The fifth question that has to be addressed is "what is the definition of each Viewpoint in terms of the identified concepts?" When following PaDRe, this is covered by the 'define viewpoint definition' task of the 'Viewpoint Definition Process'.

Essentially, the pattern definer is identifying the concepts (Ontology Elements and Ontology Relationships) that can appear on each Viewpoint and is capturing them as a FAF Viewpoint Definition View (VDV).

Each Viewpoint will show a subset of the Ontology Elements and Ontology Relationships that appear on the ODV. What concepts does each Viewpoint show? Well, exactly those Ontology Elements and Ontology Relationships grouped on the ODV that was marked up in order to help identify the Viewpoints. This is why keeping the marked up ODV is so useful. Essentially in identifying the Viewpoints you have already identified the concepts that appear on each.

So, take the marked up ODV and create VDVs for each of the Viewpoints identified. Add the Viewpoint to its relevant VDV and then add the relevant Ontology Elements and Ontology Relationships for that Viewpoint as indicated by the marked-up ODV. Use relationships such as SysML *composition* or *aggregation* to show the make-up of each Viewpoint. The VDV for the Element Structure Viewpoint is reproduced below in Figure 19.16.

Think carefully about multiplicities on these *compositions* and *aggregations*. A multiplicity of '1' or '1..*', for example, will mean that an Ontology Element is mandatory on a Viewpoint. For example, in Figure 19.16 you can see that an 'Element' is made up of zero or more 'Property'. This means that an ESV can show the Properties that are part of an Element but does not have to.

Remember, a Viewpoint can only show those Ontology Elements and Ontology Relationships that appear on the ODV. If you find that you have concepts that

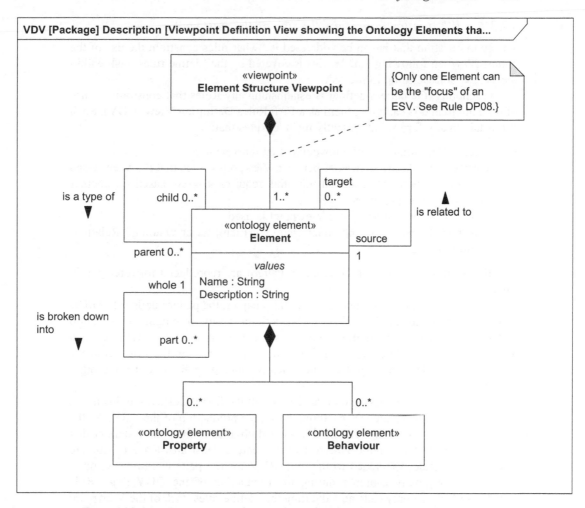

*Figure 19.16   Viewpoint Definition View for the Element Structure Viewpoint*

you want to add that are not on the ODV, then it is necessary to revisit the ODV (and in fact, the AFCV) and go through questions 1–4 again.

In summary, take the marked-up ODV used to identify Viewpoints and create a VDV for each Viewpoint. As a starting point, each VDV will contain the Ontology Elements and Ontology Relationships that were grouped for that Viewpoint on the marked-up ODV. Think about which concepts are mandatory for a Viewpoint. If missing concepts are spotted, add them to the ODV and start again by revisiting the AFCV and re-asking question 1.

Although the PaDRe 'Viewpoint Definition Process' says to create a VCV and then a VDV for each Viewpoint in turn, you may find it easier to do all the VDVs in one hit following the creating of all the VCVs. Whichever works best for you is fine.

### 19.3.6    *What rules constrain the use of the pattern?*

The sixth question that has to be addressed is "what rules constrain the use of the pattern?" When following PaDRe, this is covered by the 'define rules' task within the 'Viewpoint Definition Process'.

Essentially, the pattern definer is establishing any Rules that constrain the use of the Pattern and is capturing them as a FAF Rules Definition View (RDV). Such Rules take many forms, but typically include Rules that:

- Specify a minimum set of Viewpoints that must be used.
- Specify particular dependencies between Viewpoints, such that if a View based on one Viewpoint is produced then this requires a View based on another Viewpoint to be produced.
- Constrain the way a particular Viewpoint is used.
- Constrain the way that individual Ontology Elements or Ontology Relationships are used.

The Rules established for the Description Pattern are reproduced for reference in Figure 19.17.

As has been mentioned many times in this chapter, the pattern definer has to be flexible when producing the various Views defining a pattern; pragmatic execution of the PaDRe processes requires iteration and also an amount of moving between Views in an order dictated by the definer's thoughts more than the strict sequence implied by PaDRe. This is just as true when producing an RDV as for the other Views.

When defining the Description Pattern, one of the first Rules written down was 'Rule DP01' and this Rule was captured during the production of the VRV. As the two Viewpoints were being identified it seemed obvious to the pattern definer that anyone using the Description Pattern had to create at least an Element Structure View, so this Rule was added to the draft RDV for the pattern. Similarly, other Rules were originally captured during the production of the ODV (e.g. 'Rule DP04'). In fact, the only Rule added during the 'define rules' task of the Viewpoint Definition Process was 'Rule DP08'.

The Rules defined for the Description Pattern cover three of the four typical Rule types discussed above, namely:

- **Specify a minimum set of Viewpoints that must be used.** An example of this for the Description Pattern is 'Rule DP01'.
- **Constrain the way a particular Viewpoint is used.** An example of this for the Description Pattern is 'Rule DP08'.
- **Constrain the way that individual Ontology Elements or Ontology Relationships are used.** Rules 'Rule DP02' to 'Rule DP07' are all examples of this type of Rule.

The one type that is not overtly covered in the Description Pattern is **"Specify particular dependencies between Viewpoints, such that if a View based on one Viewpoint is produced then this requires a View based on another Viewpoint**

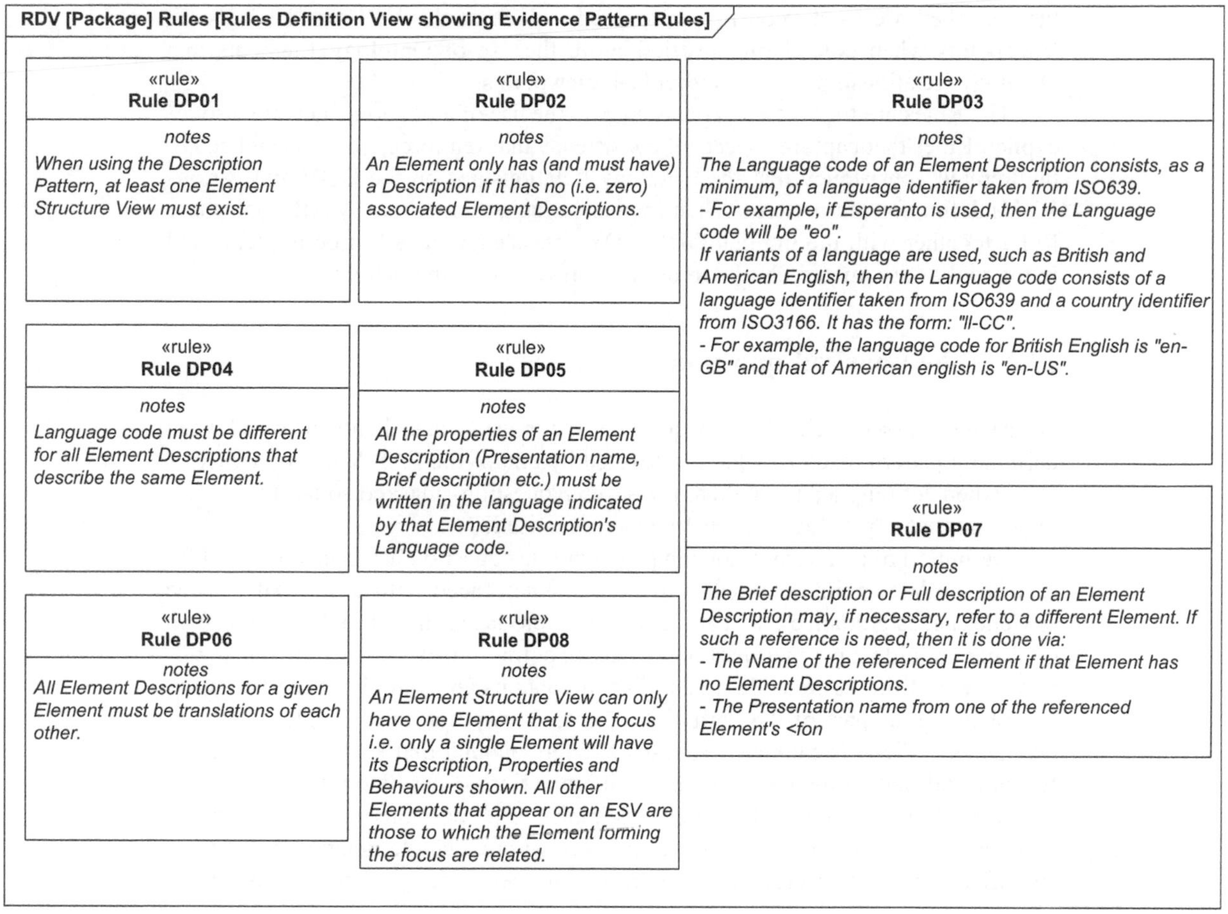

RDV [Package] Rules [Rules Definition View showing Evidence Pattern Rules]

| «rule»<br>**Rule DP01** | «rule»<br>**Rule DP02** | «rule»<br>**Rule DP03** |
|---|---|---|
| *notes*<br>*When using the Description Pattern, at least one Element Structure View must exist.* | *notes*<br>*An Element only has (and must have) a Description if it has no (i.e. zero) associated Element Descriptions.* | *notes*<br>*The Language code of an Element Description consists, as a minimum, of a language identifier taken from ISO639.*<br>*- For example, if Esperanto is used, then the Language code will be "eo".*<br>*If variants of a language are used, such as British and American English, then the Language code consists of a language identifier taken from ISO639 and a country identifier from ISO3166. It has the form: "ll-CC".*<br>*- For example, the language code for British English is "en-GB" and that of American english is "en-US".* |

| «rule»<br>**Rule DP04** | «rule»<br>**Rule DP05** |
|---|---|
| *notes*<br>*Language code must be different for all Element Descriptions that describe the same Element.* | *notes*<br>*All the properties of an Element Description (Presentation name, Brief description etc.) must be written in the language indicated by that Element Description's Language code.* |

«rule»
**Rule DP07**

*notes*
*The Brief description or Full description of an Element Description may, if necessary, refer to a different Element. If such a reference is need, then it is done via:*
*- The Name of the referenced Element if that Element has no Element Descriptions.*
*- The Presentation name from one of the referenced Element's <fon*

| «rule»<br>**Rule DP06** | «rule»<br>**Rule DP08** |
|---|---|
| *notes*<br>*All Element Descriptions for a given Element must be translations of each other.* | *notes*<br>*An Element Structure View can only have one Element that is the focus i.e. only a single Element will have its Description, Properties and Behaviours shown. All other Elements that appear on an ESV are those to which the Element forming the focus are related.* |

*Figure 19.17    Rules Definition View for the Description Pattern*

**to be produced**." Definition of a Rule of this type is left as an exercise for the reader, but think about the relationship between an Element Description Viewpoint and an Element Structure Viewpoint. Can an Element Description View be produced without producing a corresponding Element Structure View? Hint: the answer is not "yes"!

In summary, when defining Rules, think about the four types identified above and use these types, along with (typically) the ODV and VRV to think about how you want the pattern to be used. Look at relationships between Viewpoints and ask what these relationships imply in terms of a minimum set of Viewpoints that must be realised when using the pattern. Also, think about the dependencies between Viewpoints. Does realising one Viewpoint mean that you have to also realise other related Viewpoints, or can it be realised in isolation? Look at the Ontology Elements and Ontology Relationships on the ODV and think about any restrictions in the way they are used. You may also need to look at the VCVs and VDVs for Viewpoints when considering restrictions in the use of Ontology Elements and Ontology Relationships and of individual Viewpoints.

The Rules are there to add consistency to the use of a pattern. They are a set of explicit Rules that capture aspects of consistency that reinforce, build on and add to the implicit consistency that the Ontology Relationships on the ODV and various VDVs define for the pattern. When implementing patterns in a SysML tool, these Rules together with the implicit ODV/VDV consistency rules can be implemented in the tool as automated checks to ensure correct use of the pattern.

## 19.4    The unasked question

We have discussed each of the six questions that must be asked and answered when defining a pattern, each of which is covered by one of the FAF Viewpoints. However, when defining a pattern there is a seventh question, unasked so far, but no less important, namely "How is each Viewpoint realised?"

Definition of a pattern is not complete until this question has been answered for each of the Viewpoints defined in a pattern. When faced with a new pattern, the modeller needs guidance on what the Views he produces should look like and such guidance should be given as part of the pattern definition. We have done this in this book. Each of the Patterns defined in Part II contains an example of the Pattern in use, showing, as part of the definition of each Viewpoint, an example of its realisation as a View. It cannot be stressed strongly enough that such examples are a fundamental part of the definition of a pattern and must not be omitted.

It is, of course, impossible to give examples of all possible methods of realisation. Different notations are available and even within a single notation, often there are different ways of representing concepts. However, some guidance can be given:

- Use a single example. Use a single example that runs through the whole pattern. This ensures that someone using the pattern can see how the various Viewpoints work together.
- Try to use a notation targeted at your audience. This book is aimed at systems engineers and uses SysML as its modelling language, so all the examples use SysML, as far as possible, to realise Viewpoints. If defining a pattern for use in an organisation that uses the Unified Modelling Language (UML), then produce examples in UML.
- Ensure that examples have accompanying text. Don't just provide example Views with no explanation. Include a commentary on how the notation is used to represent the Ontology Elements and Ontology Relationships that appear on the View.
- Discuss obvious alternatives. If there are obvious alternatives, then these can either be discussed in the commentary on the example or given as full alternative examples. Of course, what is obvious is rather subjective, but if alternatives suggest themselves when creating an example, it is worth commenting on them.

- Remember that the same concepts (Ontology Elements and Ontology Relationships) can be realised in different ways on different Viewpoints. For example, in the "seven views" process modelling approach in chapter 7 of Holt and Perry (2013), the concept of Stakeholder Role appears on both the Requirement Context Viewpoint (RCVp) and the Stakeholder Viewpoint (SVp). Typically, an RCVp is realised as a SysML *use case diagram* with Stakeholder Roles realised as *actors*. An SVp is realised as a SysML *block definition diagram* with Stakeholder Roles realised as *blocks*.
- Ensure that all the concepts for each Viewpoint are shown. Don't miss out any concepts; the examples must provide coverage for all the concepts covered by a particular Viewpoint. If a single diagram would be too messy or complicated, provide multiple examples to illustrate all the concepts.
- Comment on Rules. Include a discussion of the conformance to Rules in the descriptive text accompanying each example.

Above all, aim to ensure that someone reading the pattern definition for the first time gets a clear understanding of how the pattern is used. Try to anticipate any questions they may have about realising Viewpoints and address such questions in the examples.

## 19.5   Summary

When defining a pattern, structure and consistency can be achieved by using the FAF and the PaDRe processes to help guide and document a pattern. The six key questions are used by the pattern definer as guidance through the application of FAF and PaDRe:

1. **What is the purpose of the pattern?** Whenever you think you might have a new pattern, ask yourself exactly what the purpose of the pattern is. Write down your thoughts on the purpose and think about what you have written some more. You are aiming to clearly articulate the purpose of the pattern. Document the purpose using an AFCV. At this point, stop and think some more. Are there too many use cases? If there are, then perhaps you have multiple patterns or even the beginning of a framework. You want each pattern to have a clearly defined and well-bounded purpose. Also, do not be afraid to capture other information as it occurs to you. It is quite common, when producing an AFCV, to start thinking about concepts that the pattern will need to support and also possible rules on how it will be used. Start to capture these thoughts on the appropriate View (for example, on an ODV or RDV).
2. **What concepts must the pattern support?** Take the AFCV and, thinking about the Use Cases on the AFCV, create an ODV containing Ontology Elements and Ontology Relationships that capture the concepts that are needed to address the Use Cases. Be prepared to begin work on the ODV while working on the AFCV; as discussed previously, this is perfectly fine and is the way that the authors work. Also, remember that the ODV forms the heart of the pattern; the Viewpoints that the pattern defines can use only (and all) those concepts

that appear on the ODV. This might mean that you will need to revisit the ODV later in the process. Perhaps, when defining a Viewpoint, new concepts will be discovered that must be added to the ODV. Or, conversely, when all Viewpoints have been defined you find that there are concepts that have not been included in any Viewpoint. If you haven't missed a Viewpoint, then these concepts should be removed from the ODV.

3. **What different ways of considering the identified concepts are required to fully understand those concepts?** Use the AFCV to identify logical groupings of Use Cases. Each such grouping is a candidate Viewpoint. Mark up the ODV to show the Ontology Elements and Ontology Relationships corresponding to these groupings. Look at the marked up ODV and think about overlap. Can a single Viewpoint achieve the same as multiple Viewpoints? Beware, however, too much combining of Viewpoints. When you are happy with the candidate Viewpoints, revise the AFCV and ODV groupings as appropriate. Name your Viewpoints and produce a VRV. At this point, keep your marked up AFCV and ODV as these annotated diagrams will prove invaluable when answering the fourth and fifth questions.

4. **What is the purpose of each Viewpoint?** Take the marked-up AFCV used to identify Viewpoints and create a VCV for each Viewpoint. As a starting point, each VCV will contain the Use Cases that were grouped for that Viewpoint on the marked-up AFCV. If necessary, expand on the Use Cases on each VCV. If missing concerns are spotted, add them to the AFCV and start again by revisiting the AFCV and re-asking question 1. Although the PaDRe Viewpoint Definition Process says to create a VCV and then a VDV for each Viewpoint in turn, you may find it easier to do all the VCVs first. Whichever works best for you is fine.

5. **What is the definition of each Viewpoint in terms of the identified concepts?** Take the marked-up ODV used to identify Viewpoints and create a VDV for each Viewpoint. As a starting point, each VDV will contain the Ontology Elements and Ontology Relationships that were grouped for that Viewpoint on the marked-up ODV. Think about which concepts are mandatory for a Viewpoint. If missing concepts are spotted, add them to the ODV and start again by revisiting the AFCV and re-asking question 1. Although the PaDRe Viewpoint Definition Process says to create a VCV and then a VDV for each Viewpoint in turn, you may find it easier to do all the VDVs in one hit following the creating of all the VCVs. Whichever works best for you is fine.

6. **What rules constrain the use of the pattern?** When defining Rules, think about the four types identified and use these types, along with (typically) the ODV and VRV to think about how you want the pattern to be used. Look at relationships between Viewpoints and ask what these relationships imply in terms of a minimum set of Viewpoints that must be realised when using the pattern. Also, think about the dependencies between Viewpoints. Does realising one Viewpoint mean that you have to also realise other related Viewpoints, or can it be realised in isolation? Look at the Ontology Elements and Ontology Relationships on the ODV and think about any restrictions in the way they are

used. You may also need to look at the VCVs and VDVs for Viewpoints when considering restrictions in the use of Ontology Elements and Ontology Relationships and of individual Viewpoints.

As well as answering the six questions and producing the various FAF Views needed to define the pattern, you must ensure that someone reading the pattern definition for the first time gets a clear understanding of how the pattern is used. Try to anticipate any questions they may have about realising Viewpoints and address such questions by creating examples of each Viewpoint in the pattern, i.e. create an example View that shows how each of the pattern Viewpoints may be realised. Use a consistent example throughout the pattern and include in your examples a discussion of the realisation method adopted.

## Reference

Holt, J. and Perry, S. *SysML for Systems Engineering; 2nd Edition: A Model-Based Approach*. London: IET Publishing; 2013.

*Chapter 20*

# Using Patterns for model assessment

## 20.1 Introduction

Patterns can be used in many ways, one of these is as an assessment tool considering the extent of existing Frameworks. This chapter will discuss the practice of model assessment by applying a simple set of metrics to different aspects of a Framework.

This chapter will provide two example assessments before describing the application of the Epoch Pattern which has been used to carry them out. The examples are not defined using the FAF as this is likely to make the assessment too easy to deliver and will not show the benefit of applying assessments to any model. The examples will take an existing ontology from a partially defined Framework and a set of Viewpoints from another partial Framework.

In each assessment a number of Patterns from Part 2 will be selected and compared with the example model, the results of this comparative assessment and discussion of the meaning and effect will follow.

## 20.2 Assessing Framework ontologies

When assessing a Framework we aim to look at all of the FAF views. However, often the complete set of views isn't available as is the case with ToGAF where an ontology is provided. This section focuses on recognising Patterns within existing frameworks and assessing the Framework completeness based on Pattern-based comparison.

The Ontology in this section has been defined purely for the purpose of this example but is representative of a number of existing Ontologies which have been defined in industry.

Figure 20.1 shows an Ontology with 'Requirement' which deliver 'Goal' and 'Solution' which answer 'Requirement'. 'Solution' is then made up of 'Process' and 'Agent' which delivers 'Function'. 'Agent' interfaces with each other and may be delivered by a 'System' or 'Role'. Each 'Interface' is made up of 'Protocol', 'Information', 'Owner' and 'Physical' aspects.

From a brief review of the available Patterns described in Part 2 there are likely to be three Patterns which can be applied to this ontology these are Context,

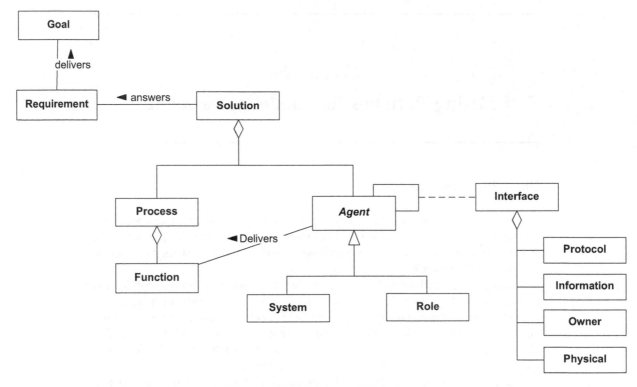

*Figure 20.1    Framework Ontology*

Interface and Description Patterns. These Patterns have been selected based on a knowledge of the purpose of the ontology in Figure 20.1 and the purposes of each of the Patterns, a fuller analysis could compare the requirements for the Ontology with Pattern aims to identify overlaps. However, in this case the requirements for the Ontology have not been formalised, as it has not been defined using the FAF, this means personal knowledge or interpretation is required reducing the rigour of the assessment.

The assessment has been carried out by taking each Pattern in turn comparing the concepts in the Pattern with those in the Ontology and comparing the relationships in the Pattern with those on the Ontology (Figure 20.1).

Firstly we will look at the way the Context Pattern maps, Figure 20.2, onto the Ontology in Figure 20.1. Considering the 'Requirement' as the 'Focus' of a 'Context description' and the 'Goal' as a 'Context' a mapping between these two Ontologies can be formed. A qualitative analysis of this mapping may show:

- 40% of concepts delivered
- 0% of relationships delivered

These metrics suggest a medium strength mapping to the concepts in the Ontology and a weak mapping to the relationships this weak mapping suggests a review of

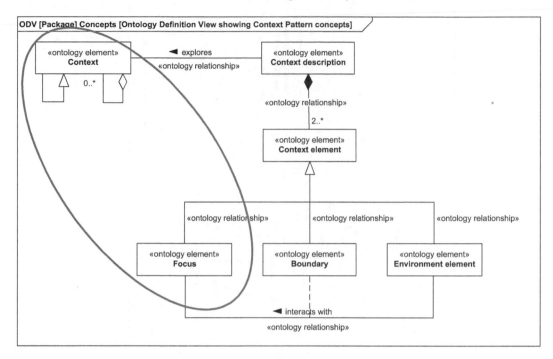

*Figure 20.2   ODV – Context Pattern*

the initial Ontology would be helpful as there is likely to be the need to extend the 'Requirement' related aspects in Figure 20.1.

Considering the Ontology from the Interface Pattern using the 'Agent' from Figure 20.1 to map to the 'System Element', the 'Interface' concept from Figure 20.1 to map to the 'Interface' Figure 20.3 and the component parts of the interface in Figure 20.1 to map to 'Interface definition', 'Port' and 'Protocol' in Figure 20.3. In this case the results of the mapping metrics are:

- 50% concept coverage
- 50% relationship coverage.

This mapping is much stronger than the previous one but still suggests there may be room for improvement, although the missing areas could be considered to be technically focused and may not be needed in the model, in this case the comparison for improvement must also consider the context of the Framework.

The Description Pattern shown in Figure 20.4 should apply to all elements in the Ontology in Figure 20.1, however, the local description concept is not defined also only a limited set of relationships are shown the results of the metrics would be:

- 50% concept coverage
- 60% relationship coverage.

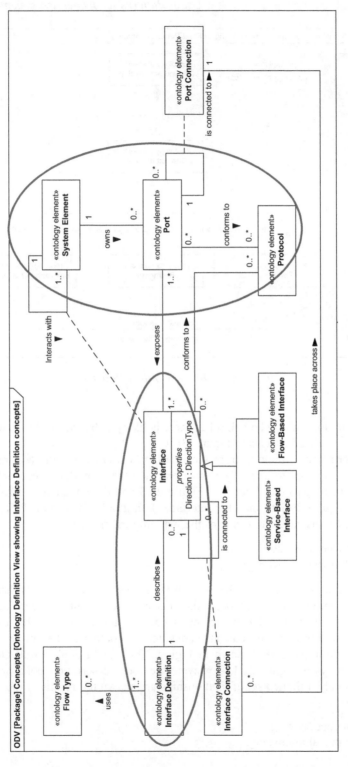

*Figure 20.3  ODV – Interface Pattern*

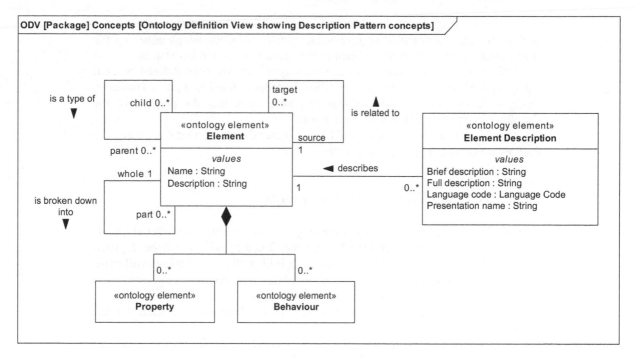

*Figure 20.4   ODV – Description Pattern*

As the Description Pattern applies to all elements on the ontology in Figure 20.1 we need to consider this result in a slightly different way, in this case the main point of the pattern is the concept of the 'Element Description' and the relationships which can be made between 'Elements'.

The Ontology in Figure 20.1 does not identify the 'Element Description' concept and as this is a critical element of every model it should be added in a relevant way. This addition may not be as simple as it sounds as before the Ontology can be updated we must check that its own purpose is still fulfilled once the aspect has been added. In this case the purpose of the ontology is to represent the implemented system and although having a description is key the element description would be out of place on a delivered system based ontology. It is likely that to solve this issue a wider consideration of the programme Ontology, rather than the design, should be developed.

### 20.2.1   Ontology assessment summary

The Ontology in Figure 20.1 although believed by its author to cover the majority of the needs of the work it was developed for. That being said it still has possibilities for change as in many cases the overlap between the Ontology and the Patterns identified for comparison equate to approximately 50% overlap.

Possible changes include additional concepts to further define specific areas such as the consideration of requirements, updates to relationships either by the way existing relationships are defined or the addition of new relationships.

This example has shown that an Ontology which has been defined without using the FAF can be assessed against the Patterns defined in Part 2. However, changes should not be made without first being sure that the purpose of the Ontology in question will not be undermined.

Finally and most obviously would be to further define the Ontology using the FAF views or as a framework suggested in Part 3.

## 20.3    Assessing Framework Viewpoints

As with the previous example here an incomplete Framework is considered, in this case the views have been defined but not the Ontology which will be depicted within each Viewpoint. Although in this case it is likely that an Ontology will exist, something similar to the example in Section 20.2, it may not have been mapped to show which aspects of the Ontology can be used in each Viewpoint. In this case we will look at the Viewpoints defined as part of the Patterns as a basis for the assessment.

In each case the metrics are split into an overall metric for the percentage of Viewpoints mapped and the relationships shown, additionally for each Pattern Viewpoint the percentage of the concerns covered by the example Viewpoint is also provided.

Figure 20.5 shows a number of Viewpoints which may be defined as part of any Architecture Framework, they are grouped into four areas, 'Requirements', 'Behaviour', 'Structure' and 'Emergent Properties'. Using a similar approach to the Ontology mapping a number of Patterns can be mapped to this set of Viewpoints, these include the Context, Description and Interface Patterns as with the previous example, but here we also consider the Analysis Pattern and the Test Pattern. The Test Pattern has been identified from the Patterns used in the Process Framework as the behavioural aspects of these views closely align to the Process Framework described in Part 3, Process modelling.

The assessment has been carried out by examining each Pattern and looking at the Viewpoints which map, based on this mapping the purpose of the Viewpoint is considered by comparing the Viewpoint Context View from the FAF with the purpose of each of the Viewpoints, in a similar situation to the Ontology example the understanding of the purpose of the Viewpoints has to be based on the Viewpoint authors interpretation as the Viewpoints have not been fully defined against the FAF.

Starting at the top of the Figure 20.5 with the 'Requirements' area we look at the Viewpoints from the Context Pattern.

From the Context Pattern the 'Context Description View', shown in Figure 20.6, maps to the 'Needs Viewpoint', in Figure 20.5, however reviewing the Viewpoint Context View for the 'Context Description Viewpoint' shows that only

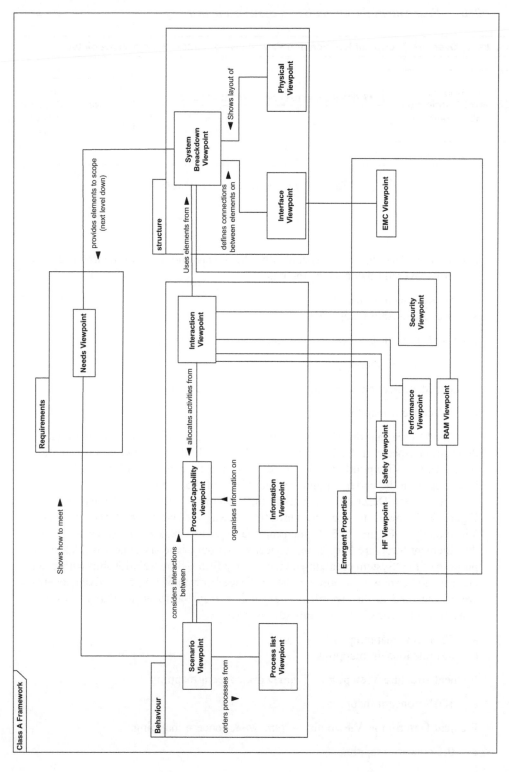

*Figure 20.5   Example views*

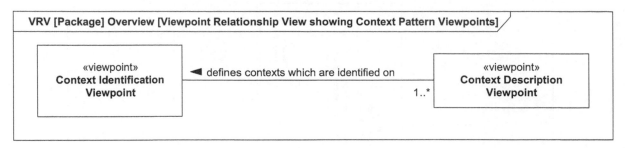

*Figure 20.6    VDV Context Pattern*

half of the Concerns map, specifically the Focus and Environmental Element concerns. Overall this provides metrics of:

- 50% Viewpoint mapping
- 0% relationship mapping.

Context Description Viewpoint – Framework concern mapping:

- 50% concern mapping.

Context Identification Viewpoint – Framework concern mapping:

- 0% concern mapping.

It is good to see that the Viewpoints in the example Framework cover the context description aspects of the Context Pattern but the clarity provided by the 'Context Identification Viewpoint' to identify those Contexts to be considered is likely to help greatly in the understanding and definition of the system.

From the Description Pattern the 'Element Structure Viewpoint', in Figure 20.7, maps to the 'System Breakdown Viewpoint' and the 'Interface Viewpoint', in Figure 20.5. Comparing the 'System Breakdown Viewpoint' with the 'Element Structure Viewpoint' concerns four out of the six concerns map to the purpose of the 'System breakdown view' with a fifth 'Allow relationship with other elements to be shown' mapping to the 'Interface Viewpoint'. As these five concerns are all included in the overall concern it is reasonable to argue that the overall concern is also satisfied. This provides metrics of:

- 50% view mapping
- 0% relationship mapping.

Element structure Viewpoint – Framework concern mapping

- 100% concern mapping.

Element Description Viewpoint – Framework concern mapping.

- 0% concern mapping.

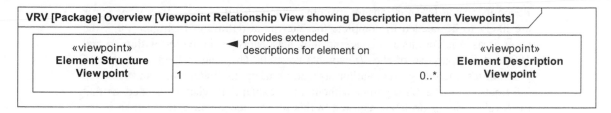

*Figure 20.7   VDV Description Pattern*

We must be clear in this case that that 100% mapping only applies to the 'Element Structure Viewpoint' as the 'Element Description Viewpoint' is not covered by the Viewpoints in Figure 20.5.

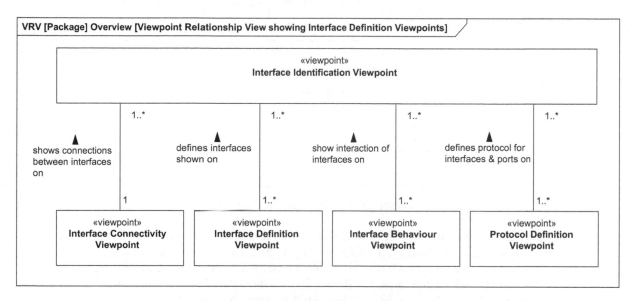

*Figure 20.8   VDV Interface Pattern*

From the Interface Pattern the 'Interface Identification Viewpoint' and the 'Interface Definition Viewpoint' in Figure 20.8 both map to the 'Interface Viewpoint' in Figure 20.5, in the first instance this suggests that there may be missing Viewpoints in Figure 20.5 as a small proportion of the Viewpoints map. The metrics show:

- 40% Viewpoint mapping
- 0% relationship mapping.

Interface Identification Viewpoint – Framework concern mapping

- 66% concern mapping.

Interface Definition Viewpoint – Framework concern mapping

- 100% concern mapping.

Recognising the 0% relationship mapping which identifies that other Viewpoints are likely to be needed to complete the Viewpoint definition in Figure 20.5, the other areas, Viewpoints and concerns, look reasonably well covered with 66% and 100% of the purpose of the Viewpoints covered. This may reflect a number of possibilities including, the author expecting users to understand and develop information on the Viewpoints without fully explaining what is needed or that the level of rigour or detail required in this set of Viewpoints is lower than in the reference Pattern. In either case before any updates are made the purpose of the Viewpoints and Framework as a whole should be considered.

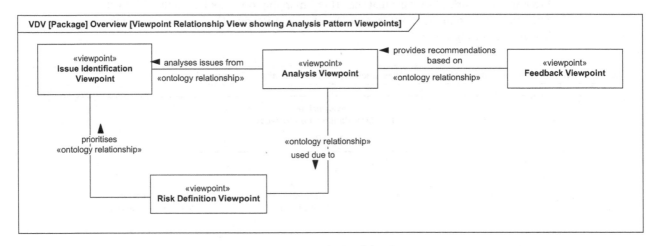

*Figure 20.9    VDV Analysis Pattern*

On the first inspection it appears that none of the Analysis Pattern Viewpoints in Figure 20.9 map to the Viewpoints in Figure 20.5, considering this as a direct comparison this appearance would be correct and the metric will score 0% for Viewpoints and relationships. On further investigation the 'Emergent properties' section of Figure 20.5 provides Viewpoints for Safety, Security, Human Factors, Performance, RAM (Reliability, Availability and Maintainability) and EMC (Electromagnetic Compatibility). These Viewpoints are each intended to cover all of the Viewpoints defined in the Analysis Pattern. In this case the detail in Figure 20.5 is more focused on multiple applications of the Pattern rather than a mapping to individual Viewpoints within the Pattern. This suggests that this area of Figure 20.5 has been delivered at a different level of abstraction in turn this suggests there is an expectation that those working with these Viewpoints understand the detail which is expected to be captured within.

The Viewpoint found in the Test Pattern, Figure 20.10, is the 'Test Behaviour Viewpoint' as a type of 'Test Case Viewpoint' which maps to the 'Interaction Viewpoint' in Figure 20.5 providing a mapping metric of:

- 14% viewpoint mapping
- 0% relationship mapping.

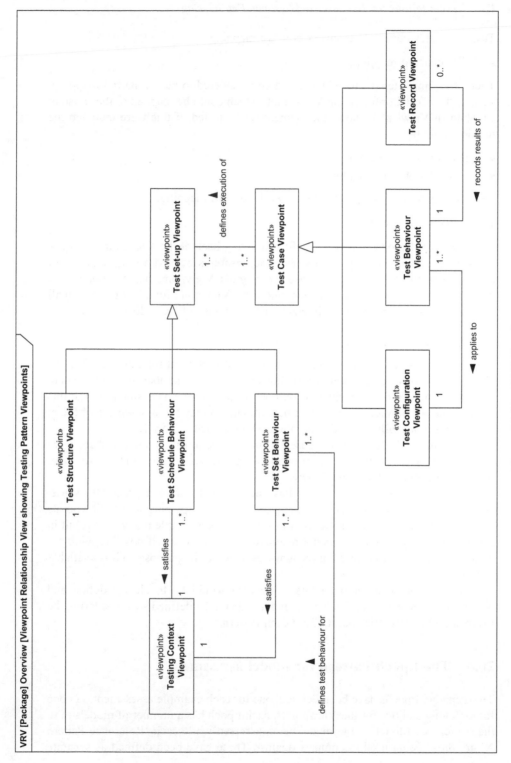

*Figure 20.10  VDV Test Pattern*

Test Case Viewpoint – Framework concern mapping

● 17% concern mapping.

There are some Viewpoints which could be considered to have a partial mapping, such as the 'Test Configuration Viewpoint' which could be mapped to the 'System Breakdown View' as it shows the elements to be tested, if this were included the metric would be:

● 28% Viewpoint mapping
● 17% relationship mapping.

Test Configuration Viewpoint – Framework concern mapping.

● 50% concern mapping.

This example demonstrates the benefit of looking back at the Viewpoint relationships once a Viewpoint has been selected to ensure associated Viewpoints do not overlap with the Pattern, in this case the original Viewpoint has a weaker 17% mapping compared with the 50% mapping of the Viewpoint found by reviewing all Viewpoints associated with the Interaction Viewpoint in Figure 20.5.

### 20.3.1   Viewpoint assessment summary

The Viewpoints defined in Figure 20.5 provides a start point for fully defining the Viewpoints to be considered, these however have a number of areas where improvements could be considered. Before considering the assessment within this section, the set of Viewpoints would benefit, and so may the assessment of them, from being more fully defined using the FAF described in Part 1.

In terms of the assessment presented in this section in each case there were options to add viewpoints, develop relationships between Viewpoints or to cover the full concerns of the mapped view.

In many cases this assessment has been carried out in a pragmatic sense, showing where the author had intent and understanding the purpose of Viewpoints from conversation. In the case where the author is not available this would result in a score of 0% as the purpose is not recorded. The discussion, if possible, with the author may be used to help them recognise and record the purpose so it is available in future.

The recommendation from this assessment would be to clearly define and record the purpose of each Viewpoint and to expand the defined set considering the mapped and missed Viewpoints from each Pattern.

## 20.4   The Epoch Pattern for model assessment

Two separate Epochs have been defined, one for each example assessment, i.e. one for Ontology and one for the Viewpoints, each Epoch has a number of models in it, the 'Reference Model' is the model being assessed while each 'Reference Pattern X' are those being used to compare against. These have been defined as separate

Epochs as the ontology and Viewpoints are not from the same Model, each could have repeated assessments in time for which new Epochs would be defined and related to these.

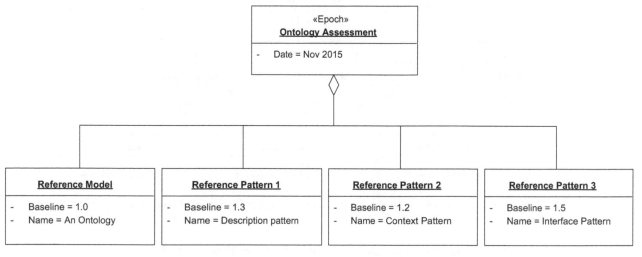

*Figure 20.11   Ontology Assessment – Epoch Definition Viewpoint*

The 'Ontology Assessment' Epoch defined in Figure 20.11 shows one 'Reference Model' called 'An Ontology' and Three 'Reference Pattern' which are the 'Description Pattern', 'Context Pattern' and 'Interface Pattern'.

For this assessment the Applicable Viewset includes the Ontology Definition View from the FAF.

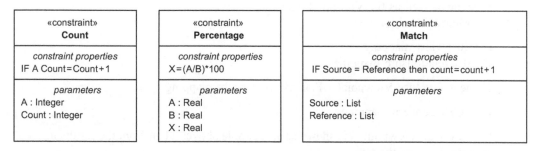

*Figure 20.12   Ontology Assessment – Metric Definition View*

The three metrics defined in Figure 20.12 are used to calculate the results for each comparison between ontologies.

- 50% concept coverage
- 60% relationship coverage.

These example results are the 'Percentage' based answers showing how well the Ontology and the Relationships map.

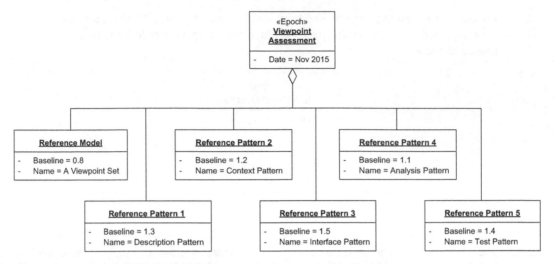

*Figure 20.13    Viewpoint Assessment – Epoch Definition Viewpoint*

The 'Viewpoint Assessment' Epoch defined in Figure 20.13 shows one 'Reference Model' called 'An Ontology' and Five 'Reference Pattern' which are the 'Description Pattern', 'Context Pattern', 'Interface Pattern', 'Analysis Pattern' and 'Test Pattern'.

For this assessment the Applicable Viewset includes the Viewpoint Relationship View and the Viewpoint Context View from the FAF.

The same metrics defined in Figure 20.12 are used in this assessment, the difference being that they are used multiple times to provide results for the concern mapping as well as the Viewpoint mapping.

Metric definition view:

- 50% view mapping
- 0% relationship mapping.

Element structure Viewpoint – Framework concern mapping

- 100% concern mapping.

Metric Usage Viewpoint – is shown as an example above, taken from the results in the main body of the text.

## Further reading

The Open Group, *The Open Group Architecture Framework*; 2011. Version 9.1. Available from http://pubs.opengroup.org/architecture/togaf9-doc/arch/ [Accessed April 2016]

*Chapter 21*

# Using Patterns for model definition

## 21.1   Introduction

Model-based Systems Engineering best practice advises the use of an Architectural Framework (AF) when developing a model of a System.

In an ideal world, a suitable ready-made AF would be available from the start of the project. Failing this, the project would start by defining its own AF using the FAF and ArchDeCon (see Holt and Perry (2013, chapter 11).

However, sometimes even this is not possible and work must start on a System model with no AF in place. However, all is not lost. There may be existing Patterns or application Frameworks that can be used. This book provides a number of enabling Patterns in Part II that can be used in the development of a System model. For example, if confronted with the need to model Interfaces, then look at the Interface Pattern. A number of application Frameworks (many discussed in Part III above) are given in Holt and Perry (2013) that can be used directly in the development of the System model. For example, need to model Requirements? Then use the ACRE Framework. Processes? Use the "seven views" Framework etc.

While such Patterns and Frameworks can be used in an ad hoc fashion, a better approach is to use them in a more controlled way, developing an AF in a piecemeal manner. This means that rather than simply using any appropriate Patterns and Frameworks in an uncontrolled and unstructured fashion, they are used to evolve an AF throughout the life cycle of a project, growing the AF and using it before growing it some more. This chapter discusses such an approach that was used by the authors over a period of nine months (and which, at the time of writing is ongoing) with a customer in the automotive domain. It shows how, starting with the ACRE Framework and using a number of enabling Patterns found in this book, a rich AF was developed that to date contains Perspectives covering Requirements, Systems, Safety & Diagnostics and Traceability.

## 21.2   The initial Framework

The initial Framework was intended to cover requirements engineering and so ACRE was used as a starting point, beginning with the standard ACRE Ontology. This is shown for reference in Figure 21.1.

Only minor changes were needed to tailor ACRE for the organisation. The organisation had the concept of Feature which replaced that of Capability, and

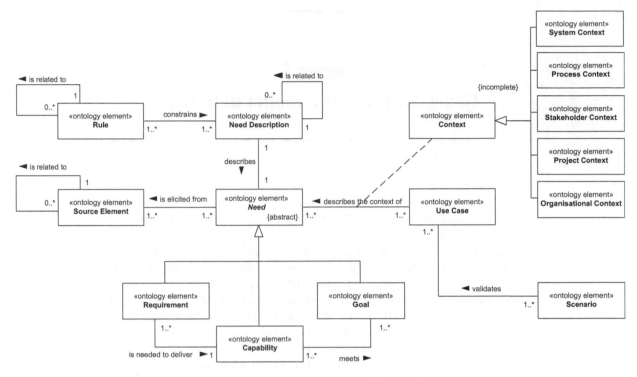

*Figure 21.1    The standard ACRE Ontology*

so Capability was renamed. They were also concerned about the quality of some of the Source Elements that they were using for the sources of Needs. A relatively cursory reading of some of the Source Elements made them realise that they would have to make a number of assumptions about Needs elicited from these Source Elements. For this reason, the concept of Assumption was added to the Ontology. The revised Ontology Definition View (ODV) is shown in Figure 21.2.

An 'Assumption' justifies one or more 'Need Description'. Any Assumption created in the model had to be confirmed as a valid Assumption; this gave a check when reviewing the model. Any Need Description with associated Assumptions could only be progressed once the Assumptions were confirmed. Such a confirmation is made through new Source Elements. If they were found to be invalid, then the related Need Descriptions were either removed from the model or revised and new Assumptions created if necessary.

The final change was not to Ontology Elements, but to the names of Viewpoints. For historical reasons, ACRE has a Requirement Description Viewpoint and a Requirement Context Viewpoint. However, these Viewpoints can show any of the types of Need (Requirement, Feature or Goal) and not just Requirements. For this reason, the opportunity was taken to rename these Viewpoints to the Need Description Viewpoint and Need Context Viewpoint. The revised Viewpoint Relationships View (VRV) is given below in Figure 21.3.

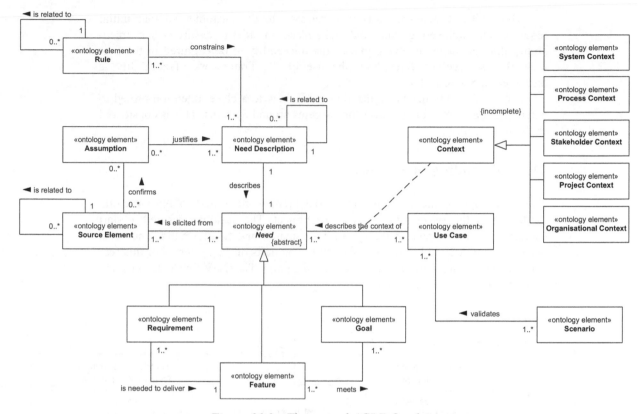

*Figure 21.2 The revised ACRE Ontology*

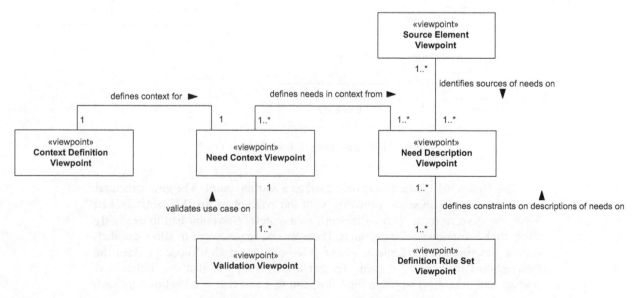

*Figure 21.3 The revised ACRE Viewpoints*

The ACRE Framework was then released to the customer so that initial requirements engineering work could take place. In order to ensure correct traceability throughout the evolving model, the traceability to be captured in the Framework also evolved throughout the use of the Framework. Traceability is discussed in Section 21.7.

As initial requirements engineering work drew to a close, attention turned to extending the Framework to cover the concepts around Systems. This is considered in the next section.

## 21.3   Extending to Systems

The next stage in the development of the Framework turned to the structural modelling of the systems involved in the project. The customer wanted to start capturing the breakdown of the system into smaller and smaller parts. They also wanted to capture the behaviour provided by the parts of the system. For this reason, the Description Pattern was a good starting point. The ODV for the Description Pattern is shown in Figure 21.4.

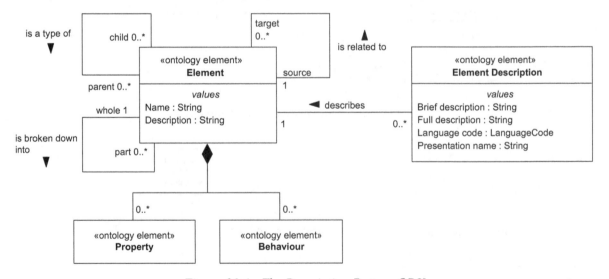

*Figure 21.4   The Description Pattern ODV*

The Description Pattern was only used as a starting point. The new structural Viewpoints were concerned primarily with the concept of an 'Element' broken down into zero or more 'part' 'Element', so any new Viewpoint had to explicitly allow such breakdown to be captured. There was no requirement to allow detailed, variant descriptions of Element, so the 'Element Description' concept from the Description Pattern was not used. Another key concern was that the 'Behaviour' concept had to be represented in the Viewpoint in a way that would allow any such

Behaviours to be traced to (see Section 21.7 for a discussion of traceability) and would also allow such Behaviours to explicitly show any messages that they send or receive. The relational concepts of 'is a type of' and 'is related to' were considered by the customer to be over-arching concepts that could, if desired, be used across any Viewpoint in the developing Framework and so would not be made an explicit part of only those Viewpoints that dealt with system structure. While discussing this with the customer, the eventual need for Viewpoints that covered interfaces was discussed, since 'is related to' could also cover interface-type relationships. Thus, this relationship, as well as being over-arching, would also be revisited when the time came to extend the Framework to cover interfaces. Therefore, a restricted version of the Description Pattern ODV was used as a starting point, as shown in Figure 21.5.

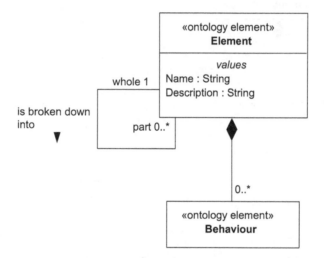

*Figure 21.5    The concepts from the Description Pattern ODV used*

The customer, as is common in most organisations, used different terms for the concept of an 'Element' depending on where it appeared in a system hierarchy. This meant that the Description Pattern ODV had to be "unwound" and the customer's terms inserted at the correct place. This was done and resulted in the ODV shown in Figure 21.6.

At the highest level of the hierarchy is the 'System' that is composed of one or more 'Subsystem' which itself is composed of one or more 'Component'. Each 'Component' is composed of one or more 'Item' and one or more 'Function', with one or more 'Function' deployed onto one or more 'Item'. Each 'Function' could send and receive one or more 'Message'. Note that although a 'Message' could be received by one or more 'Function', it could only be sent be a single 'Function'. Finally, the key links between these system concepts and those from the ACRE part of the Framework were established. An 'Function' or a 'Message' satisfies one or more 'Need Description' and each 'Need Description' has to be satisfied by zero or

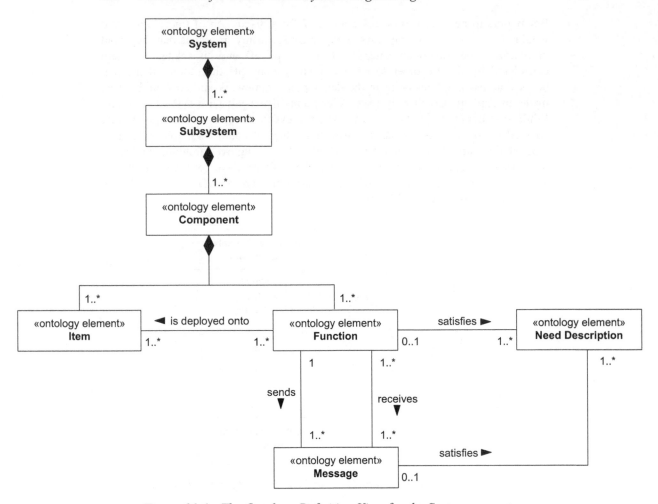

*Figure 21.6    The Ontology Definition View for the System concepts*

one 'Function' or 'Message'. This was key to the way that the customer was designing their system. Ultimately all Need Descriptions had to be satisfied by Functions and Messages; the breaking down of a System through Subsystems, Components and Items was aimed at identifying the Functions and Messages that would satisfy the Need Descriptions and deliver the system functionality.

Given that the customer used the term System, it was decided that a single 'System Structure Viewpoint' would suffice at this stage of the project. This echoes the purpose of the Element Structure Viewpoint from the Description Pattern, but uses the Ontology from Figure 21.6. The Viewpoint Definition View (VDV) for the System Structure Viewpoint is shown in Figure 21.7.

The Framework now consisted of seven Viewpoints grouped into two Perspectives, as shown in Figure 21.8.

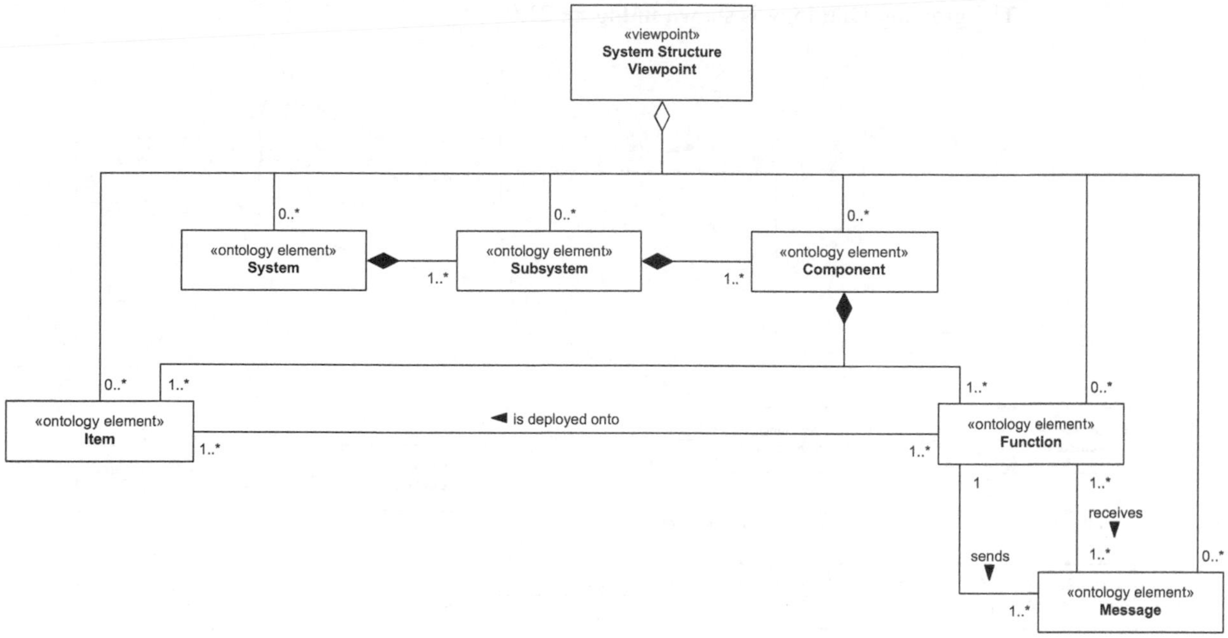

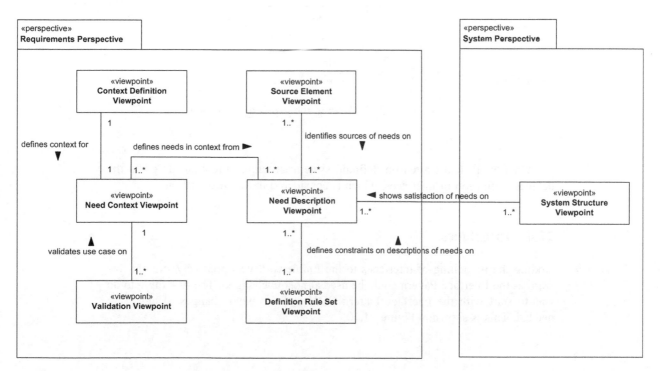

Figure 21.7   *The Viewpoint Definition View for the System Structure Viewpoint*

Figure 21.8   *VRV showing the growing Framework*

The growing Ontology is shown in Figure 21.9.

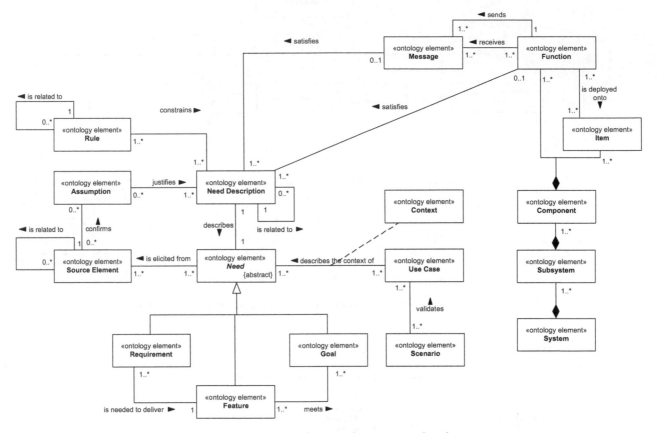

*Figure 21.9    ODV showing the growing Ontology*

Once work had started on defining system structure, attention turned to the detailed definition of interfaces, which is considered in the next section.

## 21.4   Interfaces

Adding the modelling of interfaces to the Framework was relatively straight-forward as the Interface Pattern could be used almost unchanged. The first thing to do was to start with the Interface Pattern ODV and see what changes, if any were needed. This is shown in Figure 21.10.

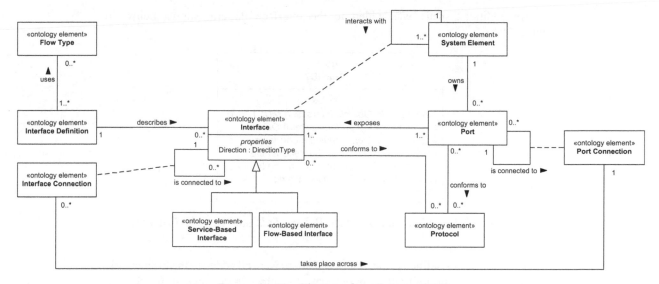

*Figure 21.10    The Interface Pattern ODV*

In the system being developed, interfaces can occur at any level in the hierarchy, so there was a need to revise the Ontology to allow this. 'System Element' was renamed 'Interfaceable Object'. This could be any of System, Subsystem, Component, Item or Function. These changes were made as shown in the revised ODV in Figure 21.11.

One additional change, shown on Figure 21.11, was making 'Message' a type of 'Flow Type'. This captured the nature of Messages in the system model, as things that flow between Functions across Flow-Based Interfaces.

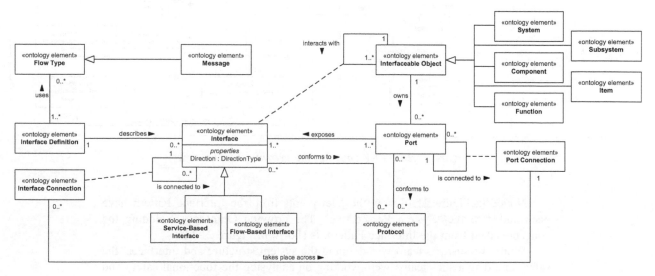

*Figure 21.11    The revised Interface Pattern ODV*

An additional Rule was added to the Interface Pattern, shown below in Figure 21.12.

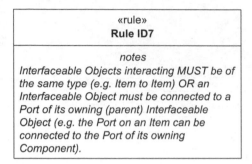

*Figure 21.12    The additional Interface Pattern Rule added to the Framework*

The Viewpoints defined by the Interface Pattern were all added to the Framework. The VDVs had to be updated as appropriate to reflect the change from 'System Element' to 'Interfaceable Object' and the fact that 'Flow Type' now had 'Message' as a sub-type. The Framework now contained the Viewpoints shown in Figure 21.13.

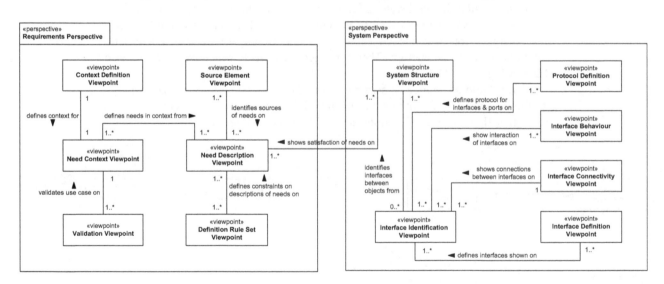

*Figure 21.13    VRV showing the growing Framework*

Notice in Figure 21.13 how the Viewpoints from the Interface Pattern have been added to the 'System Perspective'. The growing Ontology, including the concepts taken from the Interface Pattern, is shown in Figure 21.14.

While the systems team was defining the system structure and interfaces, the safety and diagnostics teams were working on analysing the functional safety and

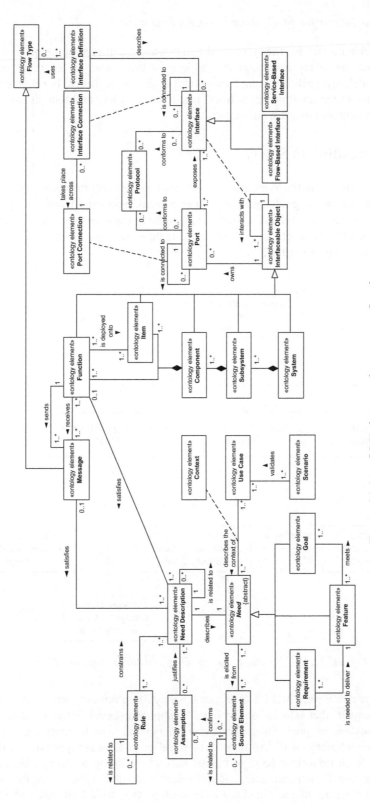

*Figure 21.14   ODV showing the growing Ontology*

functional diagnostics for the system. Although using traditional techniques, it was decided that the safety and diagnostic aspects of their work be captured in the SysML model, primarily for completeness and ease of traceability. For this reason, it was decided to extend the Framework, first to cover safety and then diagnostics. Safety is considered in the next section.

## 21.5   Extending to Safety

The organisation had the concepts of functional safety requirements and safety goals, with the safety goals applying at the System, Subsystem and Component level of the system hierarchy. The first step, therefore, was to extend the concepts in the Requirements Perspective to cover these additional concepts. This was straight-forward, as shown in Figure 21.15.

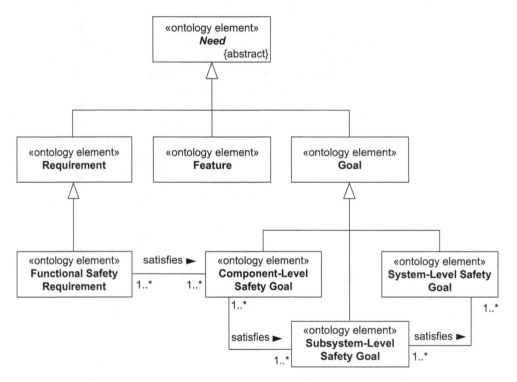

*Figure 21.15   Partial ODV showing additional types of Need*

As shown in Figure 21.15, a 'Functional Safety Requirement' was simply a type of 'Requirement' that satisfies one or more 'Component-Level Safety Goal'. These in turn satisfy one or more 'Subsystem-Level Safety Goal' which satisfy one or more 'System-Level Safety Goal'. All three of these safety goals were types of 'Goal'.

There is an important point to note here: without doing anything else to the Framework, the Functional Safety Requirement and three types of Safety Goal immediately get all of the Ontology Relationships that Requirements and Goals have through inheritance. They can be used on the same Viewpoints from the Requirements Perspective as can "vanilla" Requirements and Goals. This may seem obvious to some readers, but probably not to all, and is one of the key strengths of modelling concepts and looking for connections between new concepts and established ones. If such connections are found, such as here realising that Functional Safety Requirements are just special types of Requirement and that the three types of Safety Goal are just special types of Goal, then all of the reuse that comes in through relationships such as inheritance immediately gives a much richer and more powerful Framework without requiring any additional work.

Next the key safety concepts were captured as shown in Figure 21.16.

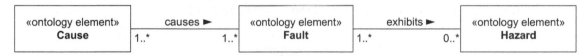

*Figure 21.16   The key safety concepts*

As shown in Figure 21.16, one or more 'Cause' causes one or more 'Fault'. One or more 'Fault' exhibits zero or more 'Hazard'. These three concepts, together with those of Functional Safety Requirement and the three types of Safety Goal, are directly related to the system structural concepts from the System Perspective. These relations are shown in Figure 21.17.

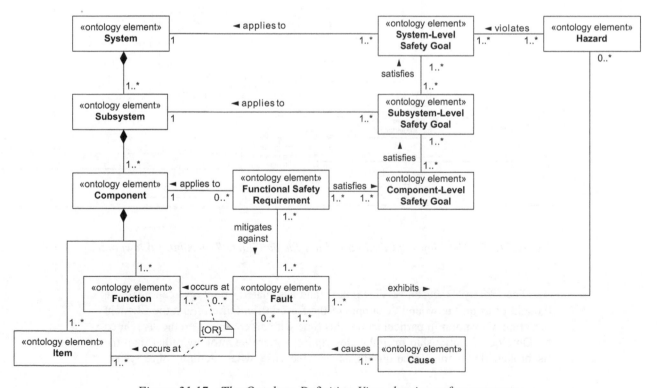

*Figure 21.17   The Ontology Definition View showing safety concepts*

Looking at Figure 21.17, it can be seen that zero or more 'Functional Safety Requirement' applies to one 'Component' and one or more 'Functional Safety Requirement' satisfies one or more 'Component-Level Safety Goal'. One or more 'Component-Level Safety Goal' satisfies one or more 'Subsystem-Level Safety Goal' which in turn satisfies one or more 'System-Level Safety Goal'. The 'Subsystem-Level Safety Goal' and 'System-Level Safety Goal' apply to a single 'Subsystem' and 'System', respectively. Note that a 'Component-Level Safety Goal' does not directly apply to a 'Component', but does so through a 'Functional Safety Requirement'. One or more 'Functional Safety Requirement' mitigates against one or more 'Fault' and one or more 'Hazard' violates one or more 'System-Level Safety Goal'. The Ontology in Figure 21.17 provided a direct definition for a 'Safety Traceability Viewpoint', shown in Figure 21.18.

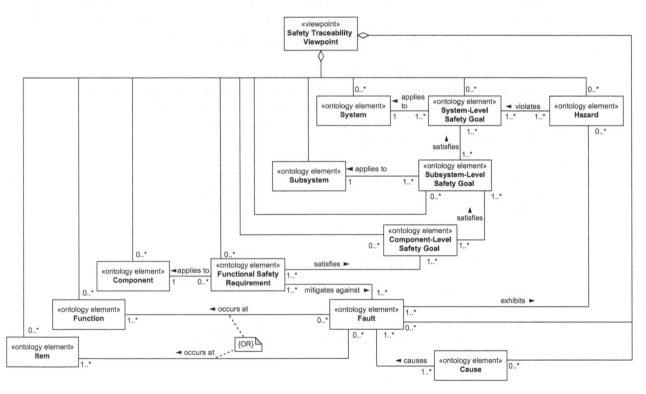

*Figure 21.18   The Viewpoint Definition View for the Safety Traceability Viewpoint*

The 'Safety Traceability Viewpoint' (and associated Ontology) can again be thought of as an "unwound" example of the Description Pattern (and the Element Structure Viewpoint in particular) but this time with an emphasis on the 'is related to' Ontology Relationship from the Description Pattern's Ontology rather than the 'is broken down into' relationship as was the case when defining the System

Structure Viewpoint. The 'Safety Traceability Viewpoint' was added to the Framework in a new 'Safety & Diagnostics Perspective', giving a new version of the Framework as shown in Figure 21.19.

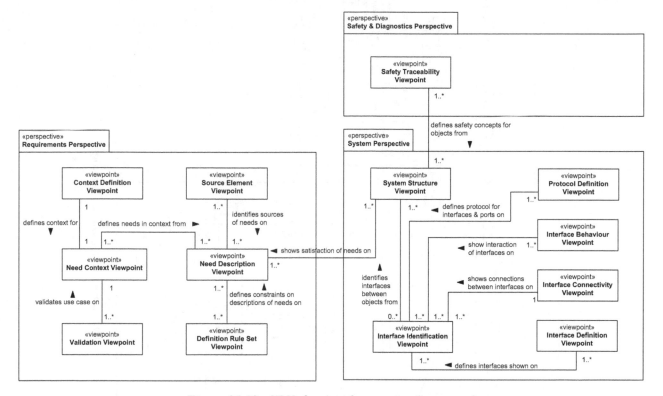

*Figure 21.19    VRV showing the growing Framework*

The concepts from Figure 21.17 are shown as part of the growing Ontology in Figure 21.20.

Once the Safety Traceability Viewpoint had been released, the Framework was then extended to cover diagnostics concepts, as discussed in the following section.

## 21.6    Extending to diagnostics

The concepts relating to diagnostics were very similar to those relating to safety. The organisation had the concepts of functional diagnostic requirements and diagnostic goals, with the diagnostic goals applying at the System, Subsystem and Component level of the system hierarchy. The first step, therefore, was to extend the concepts in the Requirements Perspective to cover these additional concepts. This was straight-forward, as shown Figure 21.21.

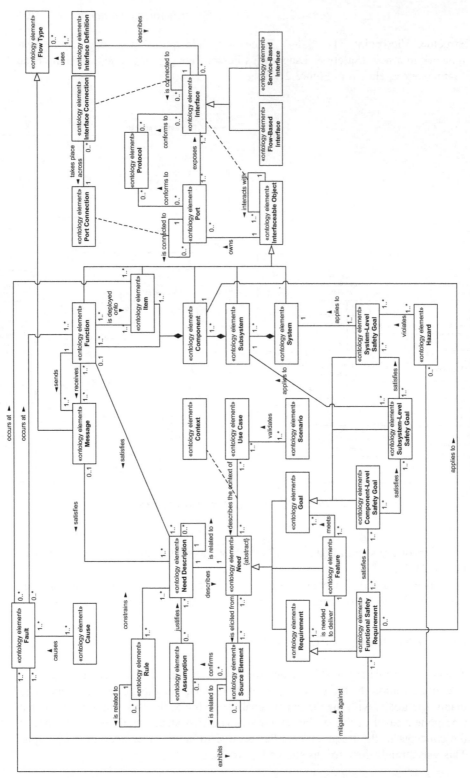

*Figure 21.20   ODV showing the growing Ontology*

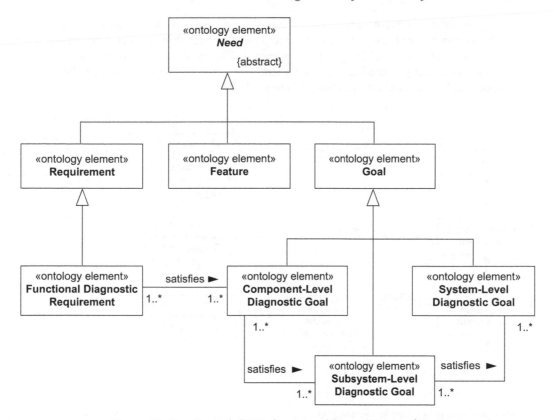

*Figure 21.21 Partial ODV showing additional types of Need*

As shown in Figure 21.21, a 'Functional Diagnostic Requirement' was simply a type of 'Requirement' that satisfies one or more 'Component-Level Diagnostic Goal'. These in turn satisfy one or more 'Subsystem-Level Diagnostic Goal' which satisfy one or more 'System-Level Diagnostic Goal'. All three of these diagnostic goals were types of 'Goal'.

As touched on above, these concepts mirrored those for safety and so, unsurprisingly, Figures 21.15 and 21.21 have exactly the same structure. This continues with the definition of diagnostic concepts shown in Figure 21.22, where 'Cause' and 'Fault' are reused from the safety concepts.

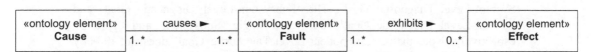

*Figure 21.22 The key diagnostic concepts*

As shown in Figure 21.22, one or more 'Cause' causes one or more 'Fault'. One or more 'Fault' exhibits zero or more 'Effect'. These three concepts, together with those of Functional Diagnostic Requirement and the three types of Diagnostic Goal, are directly related to the system structural concepts from the System Perspective. These relations are shown in Figure 21.23.

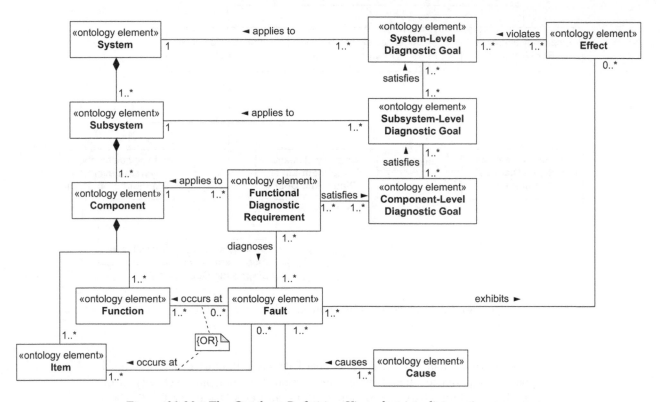

*Figure 21.23   The Ontology Definition View showing diagnostic concepts*

Again, the structure of Figure 21.23 is exactly the same as Figure 21.17 due to the mirrored concepts between safety and diagnostics. It can be seen that zero or more 'Functional Diagnostic Requirement' applies to one 'Component' and one or more 'Functional Diagnostic Requirement' satisfies one or more 'Component-Level Diagnostic Goal'. One or more 'Component-Level Diagnostic Goal' satisfies one or more 'Subsystem-Level Diagnostic Goal' which in turn satisfies one or more 'System-Level Diagnostic Goal'. The 'Subsystem-Level Diagnostic Goal' and 'System-Level Diagnostic Goal' apply to a single 'Subsystem' and 'System', respectively. Note that a 'Component-Level Diagnostic Goal' does not directly apply to a 'Component', but does so through a 'Functional Diagnostic Requirement'. One or more 'Functional Diagnostic Requirement' diagnoses one or more 'Fault' and one or more 'Effect' violates one or more 'System-Level Diagnostic

Goal'. The Ontology in Figure 21.23 provided a direct definition for a 'Diagnostics Traceability Viewpoint', shown in Figure 21.24.

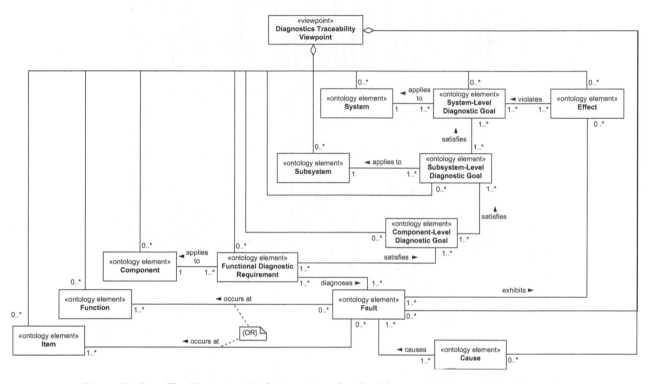

*Figure 21.24 The Viewpoint Definition View for the Diagnostics Traceability Viewpoint*

The 'Diagnostic Traceability Viewpoint' (and associated Ontology) is again an "unwound" example of the Description. The 'Diagnostics Traceability Viewpoint' was added to the 'Safety & Diagnostics Perspective', giving a new version of the Framework as shown in Figure 21.25.

The concepts from Figure 21.23 are shown as part of the growing Ontology in Figure 21.26.

The final addition to the Framework covers traceability. This is discussed in the following section.

## 21.7 Traceability

From the outset, establishing rigorous traceability was seen as essential. For this project, one of the key drivers was automation: from day one of the project it was intended that review checks be implemented in the SysML tool to create the model. Many of these were to check and report on failure to adhere to the Rules defined for the Framework (i.e. the Rules that come with ACRE and defined as necessary for additional Viewpoints). Other checks would enforce the Ontology Relationships

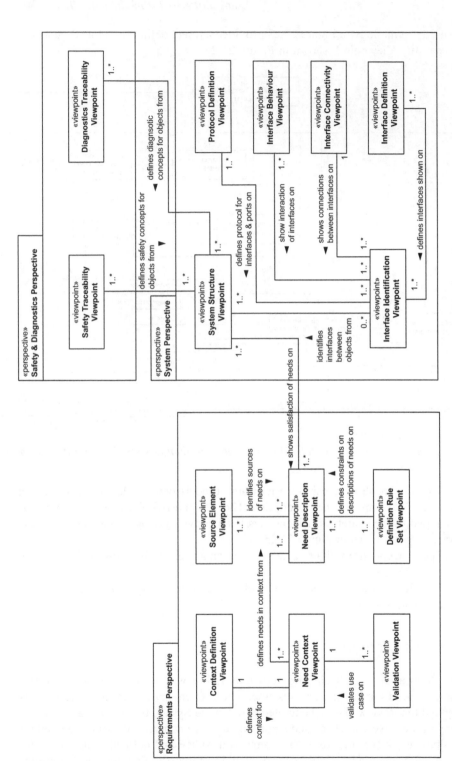

*Figure 21.25   VRV showing the growing Framework*

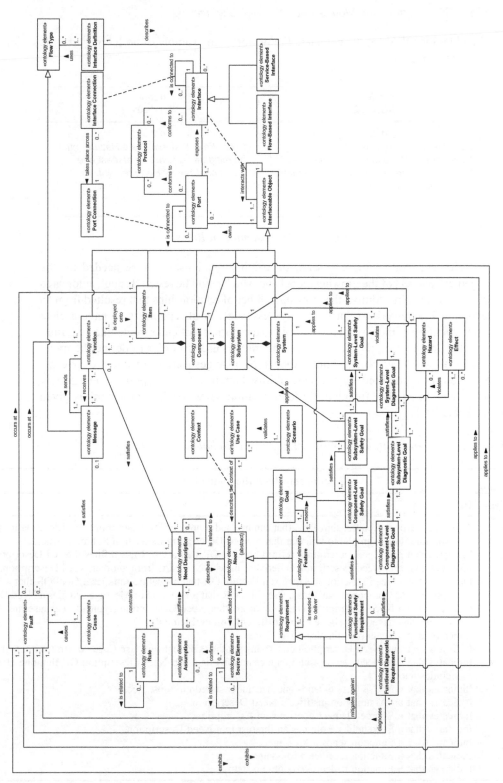

*Figure 21.26  ODV showing the growing Ontology*

found on the ODV for the complete Framework. An example of both types of Rule is given in Figure 21.27.

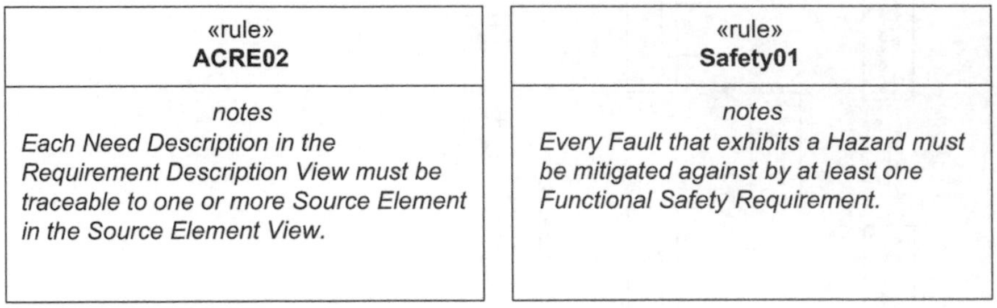

| «rule»<br>**ACRE02** | «rule»<br>**Safety01** |
|---|---|
| *notes*<br>*Each Need Description in the Requirement Description View must be traceable to one or more Source Element in the Source Element View.* | *notes*<br>*Every Fault that exhibits a Hazard must be mitigated against by at least one Functional Safety Requirement.* |

*Figure 21.27   Example of Rules*

However, other, more general, traceability relationships were needed to "tie together" aspects of the growing model and which could be tested through automated review checks. In addition, the growing number of relationships that resulted from an evolving Framework meant that it was essential for all engineers to have a clear understanding of the type of relationship to use during particular modelling activities.

For this reason, the Traceability Pattern was used in order to define the traceability allowed and expected in the model. No changes were needed to the concepts or Viewpoints defined by the Traceability Pattern; the Viewpoints were simply used as defined. The Relationship Identification View for the Framework is given in Table 21.1.

*Table 21.1   Relationship Identification View showing (partial) Relationship Types*

| Relationship Identification View | |
|---|---|
| **Relationship Type** | **Description** |
| Trace | A general purpose Relationship Type that can be used if none of the other, more specific, Relationship Types is suitable. For example, to indicate that a Need Description traces to a Source Element. |
| Refinement | Indicates that one Need Description refines another OR that a Use Case refines a Need Description. |
| Derivation | Indicates that one Need Description is derived, in whole or part, from another Need Description. |
| Satisfaction | Indicates that a Need Description satisfies a Feature OR that a Feature satisfies a Goal OR that a System, Subsystem or Component satisfies a Use Case OR that a Function or Message satisfies a Need Description. |
| Copy | Indicates that a Need Description is a copy of another Need Description (typically because the Need Descriptions have been elicited from different Source Elements). |
| Validation | Indicates that a Scenario validates a Use Case. |
| Constraint | Indicates that one Need Description constrains another OR that one Use Case constrains another. |
| Inclusion | Indicates that one Need Description includes another as a sub-Need Description OR that one Use Case includes another. |
| Extension | Indicates that one Use Case extends another under stated circumstances. |
| Justification | Indicates that an Assumption justifies a Need Description. |
| Confirmation | Indicates that a Source Element confirms an Assumption. |
| Interest | Indicates that a Stakeholder Role has an interest in a Need Description. |
| Send | Indicates that a Function sends a Message. |
| Receive | Indicates that a Function receives a Message. |
| Deployment | Indicates that a Function is deployed onto an Item. |

The RIV in Table 21.1 shows a partial list of the Relationship Types defined in the Framework. The entries shown cover the Relationship Types for the Viewpoints from the Requirements Perspective, together with those for the System Structure Viewpoint (discussed in Section 21.3). The complete RIV shows the Relationship Types for all the Framework Viewpoints. The Traceability Identification View for the same Framework Viewpoints is shown in Table 21.2.

*Table 21.2   Traceability Identification View showing (partial) Traceable Types & Relationship Types*

| Traceability Identification View | | | |
|---|---|---|---|
| **Traceable Type (Viewpoint or Viewpoint Element)** | | **Relationship Type** | **Realisation** |
| **From** | **To** | | |
| Need Description | Source Element | Trace | «trace» dependency |
| | Need Description | Derivation | «deriveReqt» dependency |
| | | Inclusion | Nesting relationship |
| | | Constraint | «constrain» dependency |
| | | Refinement | «refine» dependency |
| | | Copy | «copy» dependency |
| Need Description (Requirement) | Need Description (Feature) | Satisfaction | «satisfy» dependency |
| Need Description (Feature) | Need Description (Feature OR Goal) | Satisfaction | «satisfy» dependency |
| Assumption | Need Description | Justification | «justify» dependency |
| Use Case | Need Description | Refinement | «refine» dependency |
| | Use Case | Constraint | «constrain» dependency |
| | | Inclusion | «include» dependency |
| | | Extension | «extend» dependency |
| System Element | Use Case | Satisfaction | «satisfy» dependency |
| Scenario (Validation View) | Use Case | Validation | «validate» dependency |
| Source Element | Assumption | Confirmation | «confirm» dependency |
| Stakeholder Role | Need Description | Interest | «validate» dependency |
| Test Case | Need Description | Verification | «is interested in» dependency |
| Need Context View | Stakeholder Role (on Context Definition Viewpoint) | Trace | «trace» dependency |
| Function | Need Description | Satisfaction | «satisfy» dependency |
| | Message | Send | «send» dependency |
| | | Receive | «receive» dependency |
| | Item | Deployment | «deploy» dependency |
| Message | Need Description | Satisfaction | «satisfy» dependency |

Again, Table 21.2 shows only partial information. The complete TIV for the Framework includes the Traceable Types and Relationship Types for all the Viewpoints of the Framework.

When using the Framework, traceability is captured on a Traceability View that is typically used alongside the six Viewpoints in the Requirements Perspective and the six Viewpoints in the System Perspective. The Safety Traceability Viewpoint and Diagnostics Traceability Viewpoint (as discussed in Sections 21.5 and 21.6) are intended to capture the safety and diagnostics aspects of the model respectively, together with appropriate traceability. The full RIV and TIV contain entries relating to these two Viewpoints. However, the Framework allows safety and diagnostic traceability to also be captured on a Traceability Viewpoint if desired.

## 21.8    Summary

Best practice advises the use of an AF when developing a model of a System. In an ideal world, a suitable ready-made AF would be available from the start of the project. Failing this, the project would start by defining its own AF using the FAF and ArchDeCon.

However, sometimes even this is not possible and an AF has to be developed in a piecemeal fashion. This was the situation that existed at the start of the project used as the case study in this chapter. Nevertheless, even such a piecemeal approach can succeed if undertaken in a systematic fashion. Look at what Frameworks or Patterns already exist and see if they can be rolled into the growing Framework. This was the approach taken with the Framework developed in this chapter. The ACRE Framework formed the nucleus of the Framework, with minimal changes needed before it could be used. The Description Pattern was used to guide the addition of Ontology Elements and a Viewpoint that captured System structure. When it came to interfaces, again there existed an off-the-shelf pattern, the Interface Pattern, that could again be used with minimal changes to add an additional five Viewpoints to the growing System Perspective. The Description Pattern, along with minor additions to the Ontology Elements relating to requirements, was used twice more to define the concepts around safety and diagnostics. Finally, the Traceability Pattern was used with no changes to add a Traceability Viewpoint. From an initial off-the-shelf Framework (ACRE) the Framework has evolved over a period of nine months to a detailed and rich Ontology, as shown in Figure 21.28.

From an initial six Viewpoints in one Perspective, the Framework (at the time of writing) now has 18 Viewpoints in four Perspectives, as shown in Figure 21.29.

Before concluding this chapter, it is worth noting two points. First, defining a Framework is only part of the story. For this to be successful the Framework has to be implemented in a modelling tool; without doing this, you are not using the full power of MBSE but simply drawing pictures. All SysML modelling tools of any worth will allow such a Framework to be implemented in the tool as a SysML profile, with greater or lesser amounts of automation possible in such an implementation. Ideally, you should use a tool that will allow the consistency of the model to be checked against the Framework. Some tools will allow a fine level of control during model creation, completely preventing the modeller from creating diagrams and elements that contradict the Framework, others are less rigorous during creation and rely on the definition of checks that are run after creation to spot mistakes. If the tool that you are using will allow no such profiling and automation, then you are using the wrong tool.

The second point that is worth emphasising is that of responsibility. Framework development works best when there are only a small number (one to three) of people developing the Framework. At the very least there has to be a technical authority for the Framework to ensure coherent and consistent development and to control the release of the Framework to its users. Failure to do this will, despite the best intentions of all involved, lead to inconsistencies in concepts and between

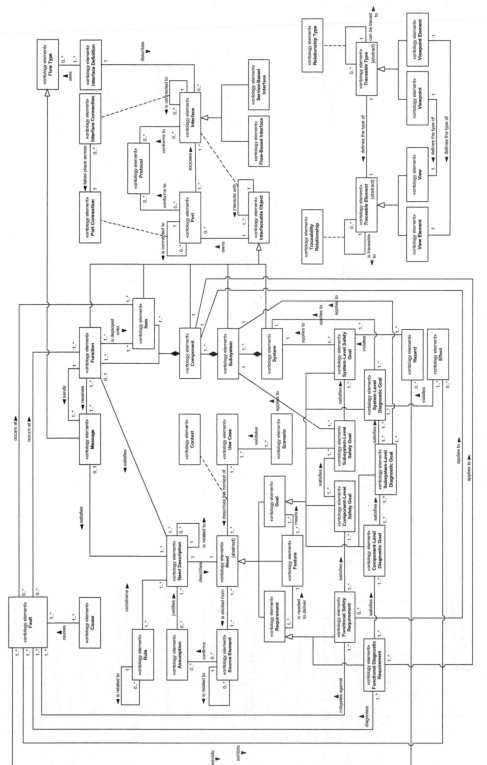

*Figure 21.28   ODV showing the complete Ontology*

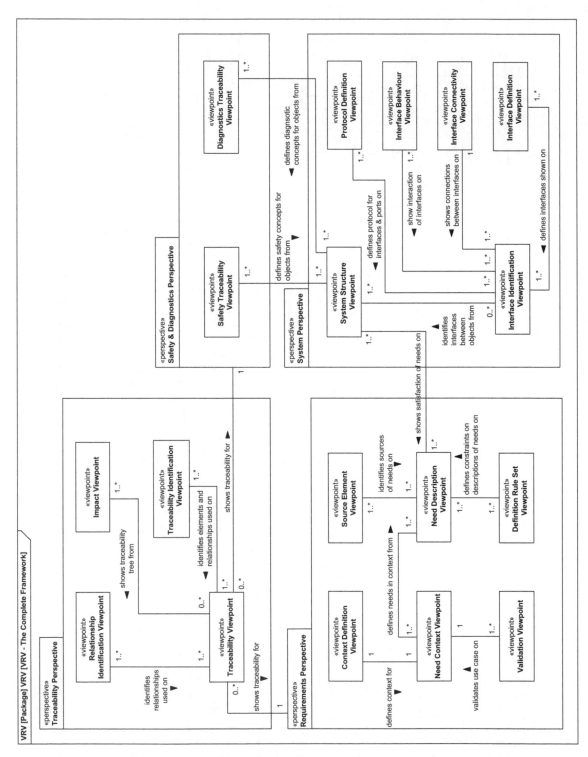

*Figure 21.29   VRV showing the complete Framework*

Viewpoints, often leads to bloat of the Framework, results (on large projects) in different parts of the project using different versions of the Framework and ultimately leads to disillusionment and abandonment of the Framework.

## Reference

Holt, J. and Perry, S. *SysML for Systems Engineering; 2nd Edition: A Model-Based Approach.*London: IET Publishing; 2013.

## Further reading

Holt, J., Perry, S. and Brownsword, M. *Model-Based Requirements Engineering.* London: IET Publishing; 2012.

*Chapter 22*

# Using Patterns for model retro-fitting

## 22.1   Introduction

Whatever the day, week or year there will be models and designs which haven't been defined or formalised using an Architecture Framework (AF). We know that Model-based Systems Engineering best practice advises the use of an AF when developing a model of a System – but this doesn't mean it will always happen.

Where organisations or industries don't have the necessary experience or maturity to carry out MBSE properly, all too often models will be developed without a Framework to guide them. In some cases these models maybe no more than unconnected representations and pictures which are described as "The design."

Even in cases where AFs are used in new projects, people often cite issues such as the AF stifling thinking or the level of rigour and rules, meaning people spend more time on thinking about the AF rather than the design. In these situations we need to find ways to bring the value of the AF without the pain which is perceived.

This chapter discusses examples of these situations, focusing on how to take unconnected/uncontrolled pictures and bringing them together using Patterns, it also addresses pragmatic ways to approach this for those who want an unconstrained approach but recognise they need a formalised output.

## 22.2   An unstructured model

The first example is an example of some of the thinking which often happens on projects and becomes so well-recognised it is considered the design. Let's be clear, there is a major difference between a diagram which is structured, consistent and cohesive as part of a Model and something which people are familiar with and therefore accept and discuss as design. This is not trying to say that the diagram is not helpful only that it is unlikely to be complete/rigorous enough to be considered a design.

Pictures in this sense are often referred to as "overviews," "architectures," "visions" and many other names. They are important as they give people a way to associate with a piece of work but they are a communication tool not a design tool.

*Figure 22.1    System Overview*

Either way the issue we are faced with here, which we often have in reality, is how to turn this image into an Architecture or Model. This is where Patterns can be of use. The first thing to do is to select a Pattern. To do this we must have some idea of the purpose of the diagram provided, Figure 22.1, and the purpose of the model to be developed. In this case the purpose of Figure 22.1 is to *provide an overview of the purpose of the project and system to be delivered*; it's irrelevant whether we believe this does its job or not (it is what we have to work with). The purpose of the Model we desire is to provide a design. This may not be the most helpful statement but is often the one we are greeted with when people have realised they have a problem understanding their design but don't know how to solve it.

The intention is to select the Pattern which best maps to both purposes. For Figure 22.1 this can be considered to be the need for the system and its hierarchy. For the Model this will be similar although we would like to add function. The selected Patterns to formalise this diagram are the Context Pattern and Description Pattern; the Context Pattern as it provides the focus of the system and the environmental elements around it and the Description Pattern as some details within the System will need further definition.

### 22.2.1   Formalising Context

Formalising Figure 22.1 using the Context Pattern provides the ability to understand the elements outside the System and those things it must deliver. It is being considered first here as it should help to clarify the purpose of Figure 22.1.

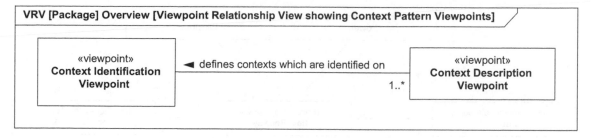

*Figure 22.2    Context Pattern Viewpoints*

The Context Pattern (Figure 22.2) reminds us that we should be defining at least two Views, one conforming to the Context Identification Viewpoint and one against the Context Description Viewpoint.

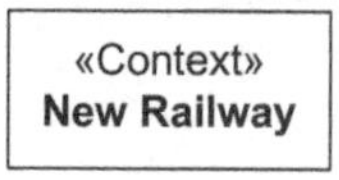

«Context»
**New Railway**

*Figure 22.3    Context Identification View – New Railway Contexts*

The Context Identification View, in Figure 22.3, shows that the 'New Railway' context should be defined which will be discussed in the following. In the meantime, the concept of a diagram with only one 'Context' on may, to some, seem nonsensical. However, it has its value – in the right circumstances it can be used to stimulate discussion, such as posing the question "should there only be one box?" and "What others could there be?" We will revisit this diagram as we progress through the formalisation as we should with all diagrams.

Figure 22.4 applies the concepts from the Context Pattern to Figure 22.1, providing the Patterns 'Environmental elements', 'Focus' and 'Boundary' based on the ideas in Figure 22.1. Again the intention here is to pick out two groups of 'Environmental Element': those things which are clearly outside the 'Boundary', such as the 'Weather', the 'Road Network' and 'Public' and those which maybe contentious such as 'Power' (based on the suggestion that a turbine could be inside the system) and 'Drainage', which could be included if this is to be an infrastructure programme. There may, in truth, be many other 'Environmental elements' but, staying true to the goal of using the Pattern to formalise the picture, these cannot be formalised here.

Taking the content of Figure 22.1 into account may well provide 'Focus' within the 'Boundary' as suggested in Figure 22.5 which expands on Figure 22.4. The 'Focus' is to 'Provide new railway' including 'Provide infrastructure', 'Provide rolling stock', 'Provide operations' and 'Provide maintenance'. These areas define the main things the railway must do whilst being constrained by 'Enable Safety', with a specific type of safety being around 'Provide safe crossing'.

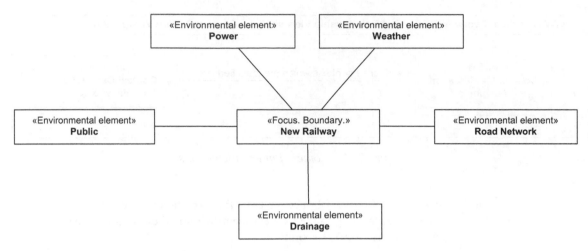

*Figure 22.4   Context Description View – Environmental elements*

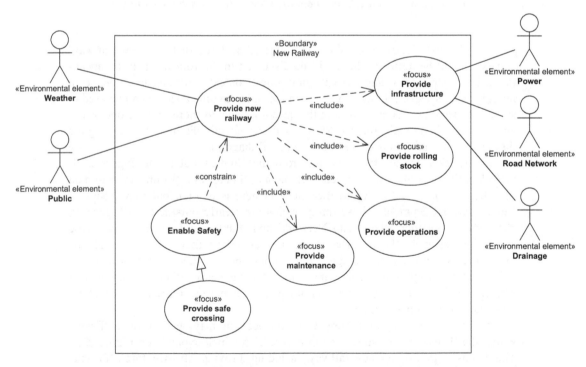

*Figure 22.5   Context Description View – System Context*

The 'Focus' have been identified by examining Figure 22.1 and highlighting those things which the author of the diagram believes to be important. This includes 'Provide Operations' due to the running train and people at the station, 'Provide maintenance' due to the depot, 'Provide rolling stock' due to the multiple representations of trains and 'Provide infrastructure' as the majority of the items on the diagram include power distribution, track, bridges and telecoms which can all be considered to be infrastructure.

The 'Environmental elements' in Figures 22.4 and 22.5 may have their own understanding of what should be provided and as such maybe added to the 'Context Identification View' in Figure 22.3 providing a possible five more 'Context' to be explored, these have been added and shown in Figure 22.6.

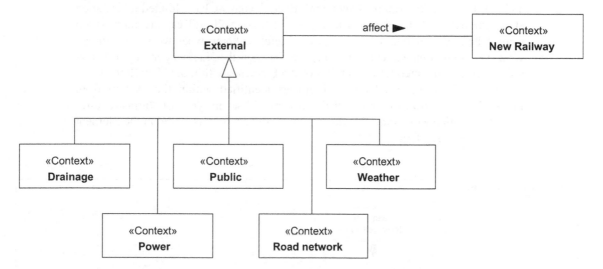

*Figure 22.6    Context Identification View – Updated*

The 'Environmental Element' from the 'Context Description View' in Figures 22.4 and 22.5 has been grouped together as special types of 'External' Context. The 'Environmental Element' has been identified as possible Contexts as they are likely to have needs associated with the new railway which can be represented within a Context. These added 'Context' are clearly only external to the 'New Railway' which they affect. In this way we can build a list of Contexts and think about how they relate to each other. Even though aspects of the 'External' Contexts may be interpreted from Figure 22.1, it is more likely that a conversation with an expert or relevant project will yield a more complete result.

## 22.2.2    Formalising Structure

Having considered the purpose of Figure 22.1 we now look at the structure provided by using the Description Pattern (Figure 22.7).

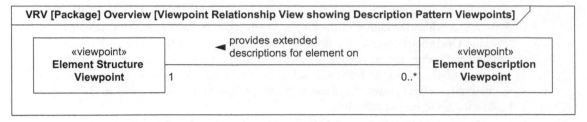

*Figure 22.7    Description Pattern Viewpoints*

The Description Pattern has two Viewpoints: the 'Element Structure Viewpoint' and the 'Element Description Viewpoint'. Even though we have limited information about the detail of the elements in Figure 22.1, we will still capture information against both Viewpoints as it may prove useful to know what we were thinking when we defined the structure. Initially we focus on the 'Element Structure Viewpoint' defining the hierarchy and types of the Elements within the New Railway.

Figure 22.8 shows all of the Elements identified within the System from Figure 22.1. It shows four parts of the system and seven types of 'Infrastructure' identified from the diagram. The description of the diagram is stored in the Element Description View, Figure 22.9.

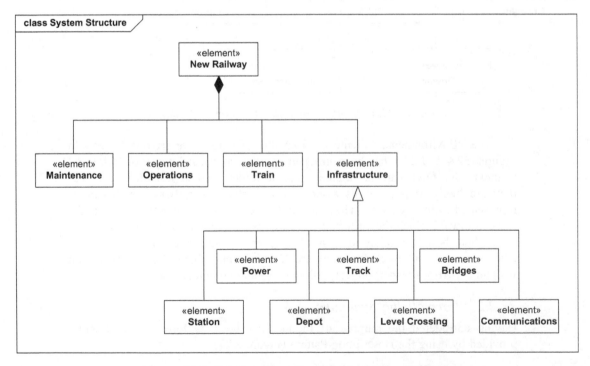

*Figure 22.8    Element Structure View – System Structure*

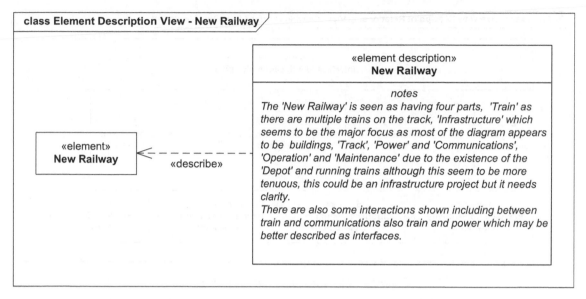

*Figure 22.9 Element Description View – New Railway Description*

The 'element description' in Figure 22.9 identifies the potential to use another Pattern, this time looking at the Interfaces identified in Figure 22.1, initially between 'Train' and 'Power' along with 'Train' and 'Communications'.

This recognition of further formalisation opportunities often occurs part way through a formalisation of a Model as the purpose and details become clear and, if relevant, we can use the Interface Pattern to formalise these connections.

## 22.2.3 Formalising Interfaces

Having identified the need to capture Interfaces we remind ourselves of the Interface Pattern.

The Interface Pattern Viewpoints, Figure 22.10, include: 'Interface Identification Viewpoint', 'Interface Connectivity Viewpoint', 'Interface Definition Viewpoint', 'Interface Behaviour Viewpoint' and 'Protocol Definition Viewpoint'. In this section of the example we will only use the 'Interface Identification Viewpoint'; the use of only a single Viewpoint means that we should be asking questions relating to the others: why haven't we used them? and what information are we missing because we haven't?

Figure 22.11 identifies the Interfaces and the Systems to which they connect using the Interface Identification Viewpoint from the Interface Pattern. The other Viewpoints are not used at this point as there isn't enough detail on Figure 22.1 to provide further definition. In fact, it could be argued that the recognition of the 'Overhead Line' and 'Pantograph' *ports* is over stating the information in Figure 22.1 and the *ports* should be left un-defined, similar to the 'Communications' to 'Train' connection, until this is properly understood. Perhaps the system may need

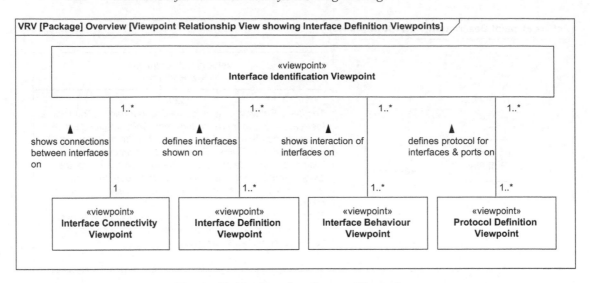

*Figure 22.10    Interface Pattern Viewpoints*

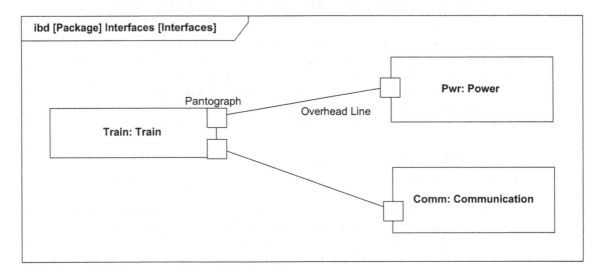

*Figure 22.11    Interface Identification View – New Railway Interfaces*

to use a different delivery method such as 3rd Rail, when the power is delivered to a train via something which looks like an additional track, rather than through an overhead cable.

We could also consider defining the Interfaces to the Environmental Elements. Although this has not been considered here, it would highlight questions around the relationship and difference between the 'Power' within the System and the 'Power' which is defined as an Environmental Element.

The key point to be taken from the discussion around Figures 22.10 and 22.11 is that they identify questions which will need to be asked and answered. Questions do not always have to be answered immediately but they do need to be managed to ensure an answer will be found.

### 22.2.4 Formalisation summary

This example has taken a rather simple picture and formalised it using three of the Patterns defined within this book: the Context, Description and Interface Patterns. This hasn't been carried out to provide a complete design as many people expect from a model, but rather it has been done to understand and enable further discussion about the topics identified within the diagram.

The formalisation of Figure 22.1 provided the recognition of additional Contexts on a Context Identification View as well as the use of an additional Pattern not originally identified for use. In turn the application of the Interface Pattern provides further questions about the number and scope of Interfaces to be captured, considering whether to include the Environmental Elements as well as the internal system ones.

Alongside the use of the Patterns, but not so easily seen from this example, the formalisation has provided a Model which has consistency between the Context, Description and Interfaces Views. This means that when one aspect of the system is questioned or developed then the effect on the others can be clearly assessed.

## 22.3 Other applications

This chapter has worked through an example based on a generic picture. This approach can also be applied to other types of picture and also to Models if there are questions around the consistency and comprehension.

Documents often contain multiple diagrams as well as large sections of text. The application of Patterns across multiple diagrams can help to understand the meaning and consistency provided by documents. This approach will highlight gaps, inconsistencies and areas of conflict should they exist. It can also be applied to the information within the text of the document by drawing out the detail and recording specific descriptions against elements. In this sense documents may include Standards, Policies, project documentation, systems documentation, minutes of meeting, etc.

Applying this approach to documentation is relatively easy as, generally, the information isn't changing. We can also apply this approach in a fast-paced design environment where the number of people who are competent in modelling and/or architecture may be limited as people will have their own specialisms. This type of situation often lends itself to an iterative approach which is beneficial as it supports the Pattern applications. The key to making this type of approach work is to allow subject matter experts space to work in the way that best suits them whilst keeping the tasks they have short and focused. This can be considered as generating raw engineering output. Once the raw output is available, the Patterns are applied to formalise the raw output and provide a formalised Model which the expert can agree and use as a basis for further work.

The benefits of this is that it can be done in short bursts allowing people to think before constraining the way they move forward with the relevant information, ensuring they work within the relevant boundaries. The formalisation can also be completed relatively quickly as there is less thinking involved if tools have been properly set up with profiles for the Patterns, enabling efficient entry of the information.

## 22.4   Summary

Although best practice advises the use of an AF, it isn't always pragmatic from the outset; often diagrams or models have already been developed and we must fit them into a Framework without destroying the author's meaning.

Using the approach in this chapter we identify Patterns which may provide relevant understanding of the diagram and use these, once populated, as a basis for further understanding and to expand to related Patterns.

In this chapter we have taken a diagram which has not been formally developed and shown that many of the concepts recognised on the diagram could be formalised without much additional work. The areas such as additional Contexts, multiple meaning for words like 'Power' and additional Interfaces where improvements have been identified and made represent issues which may have come to light on a project at a later date. As with many projects these issues are currently realised through overruns, overspends and, in the worst cases, news articles referencing the failures of the project.

The purpose of this chapter is to show that it is never too late to formalise the information which has been gathered. The formalisation provides a way for people to interact with the information in a new way and therefore identify and, if relevant, fix issues that are found.

Benefits of this approach include the ability to show a diagram in a different light which helps to identify discrepancies, whether they are related to communication issues, understanding or complexity: then highlighting questions to be asked and answered to develop the Model. When the approach is used in an iterative fashion, as within this example updating the Context Identification View and recognising that additional Patterns may be applied, it can be used as a strong tool and may be a pre-cursor to both the wider application of modelling and the use of AFs.

## Further readings

Holt, J. and Perry, S. *SysML for Systems Engineering; 2nd Edition: A Model-Based Approach*. London: IET Publishing; 2013.

Holt, J., Perry, S. and Brownsword, M. *Model-Based Requirements Engineering*. London: IET Publishing; 2012.

*Part V*
**Annex**

*Appendix A*

# Summary of SysML notation

## A.1 Introduction

This appendix provides a summary of the meta-model and notation diagrams for SysML. For each of the nine SysML diagram types, grouped into *structural* and *behavioural diagrams*, two diagrams are given:

- A partial meta-model for that diagram type.
- The notation used on that diagram type.

The same information is also given for the SysML *allocation* auxiliary construct. For full details of the SysML language readers are directed to Chapters 4–6 of "SysML for Systems Engineering" (Holt and Perry, 2013).

## A.2 Structural diagrams

This section contains diagrams for each of the five SysML *structural diagrams*:

- *Block definition diagrams* (Figures A.2 and A.3)
- *Internal block diagrams* (Figures A.4 and A.5)
- *Package diagrams* (Figures A.6 and A.7)
- *Parametric diagrams* (Figures A.8 and A.9)
- *Requirement diagrams* (Figures A.10 and A.11)

The five diagrams are shown in Figure A.1.

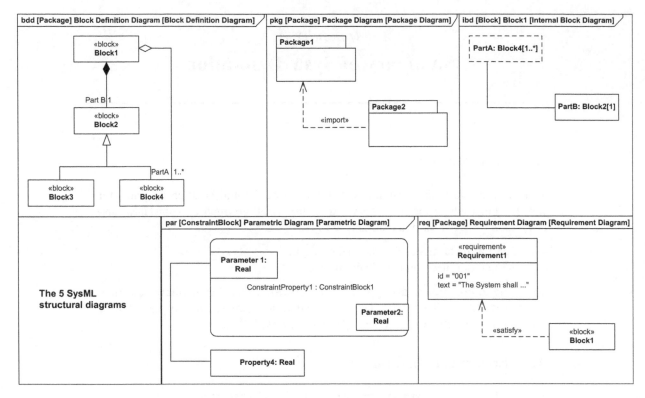

*Figure A.1    Summary of structural diagrams*

## A.2.1 Block definition diagrams

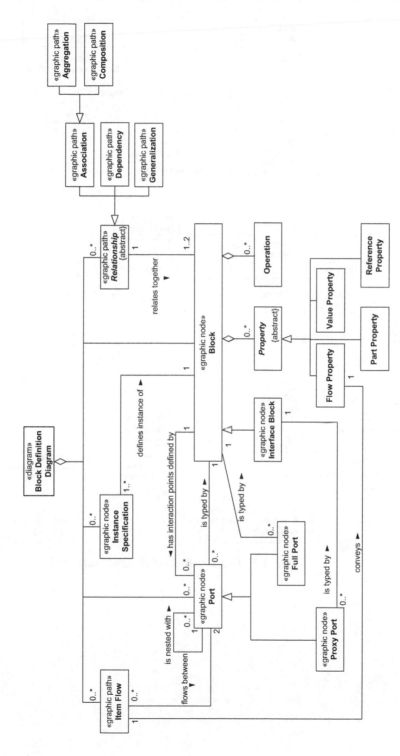

Figure A.2 Partial meta-model for block definition diagrams

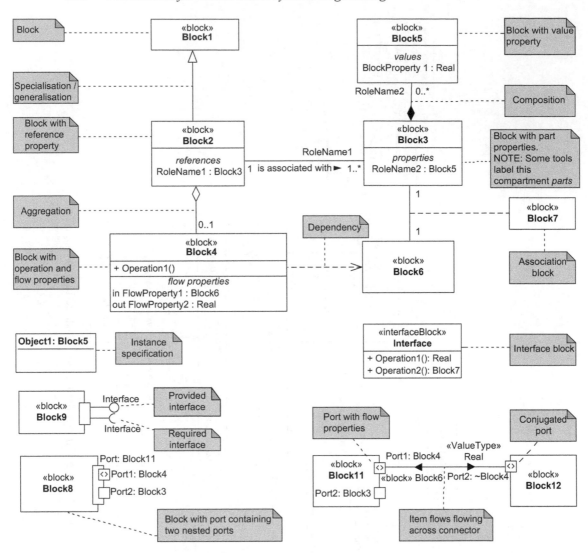

*Figure A.3*   Block definition diagram *notation*

## A.2.2 Internal block diagrams

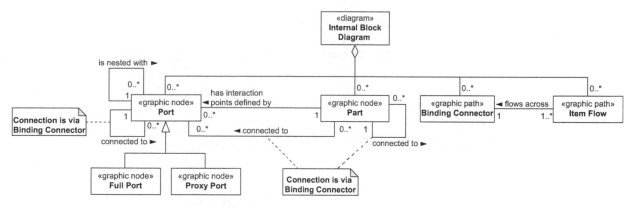

*Figure A.4 Partial meta-model for the* internal block diagram

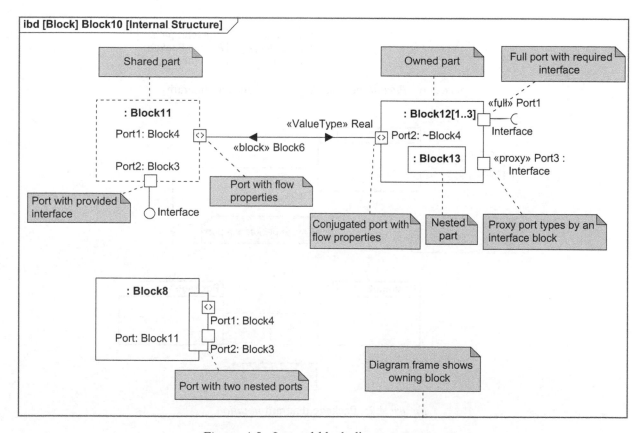

*Figure A.5* Internal block diagram *notation*

## A.2.3   *Package diagrams*

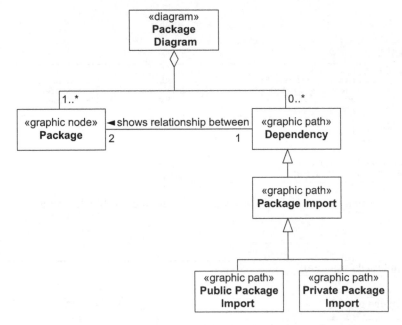

*Figure A.6   Partial meta-model for the* package diagram

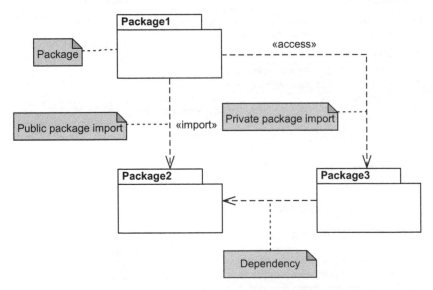

*Figure A.7   Package diagram notation*

*A.2.4   Parametric diagram*

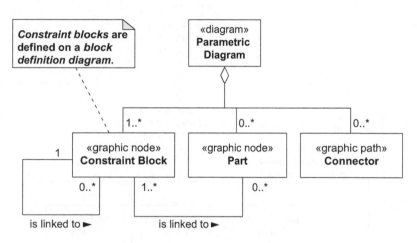

*Figure A.8   Partial meta-model for the* parametric diagram

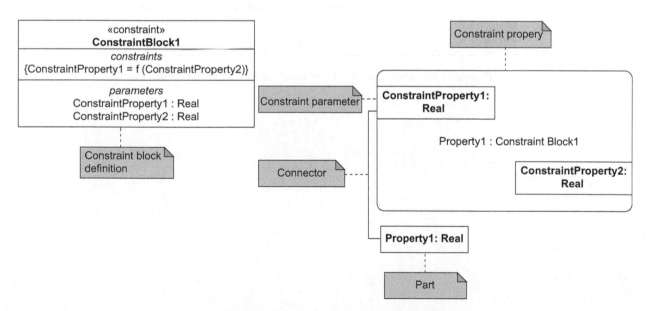

*Figure A.9*   Parametric diagram *notation*

## A.2.5    Requirement diagrams

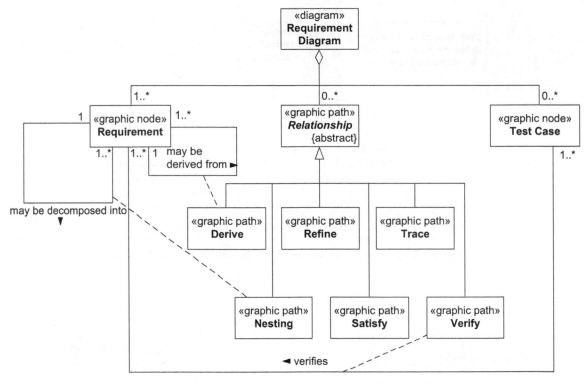

*Figure A.10    Partial meta-model for the* requirement diagram

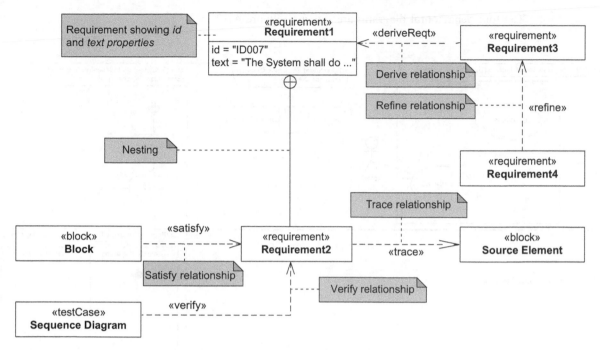

*Figure A.11*    Requirement diagram *notation*

## A.3    Behavioural diagrams

This section contains diagrams for each of the four SysML *behavioural* diagrams:

- *State machine diagrams* (Figure A.13 and A.14)
- *Sequence diagrams* (Figure A.15 and A.16)
- *Activity diagrams* (Figure A.17 and A.18)
- *Use case diagrams* (Figure A.19 and A.20)

The four behavioural diagrams are shown in Figure A.12.

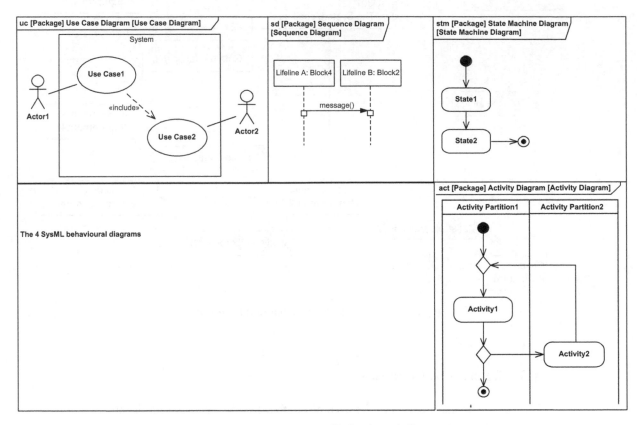

*Figure A.12    Summary of* behavioural diagrams

### A.3.1   State machine diagrams

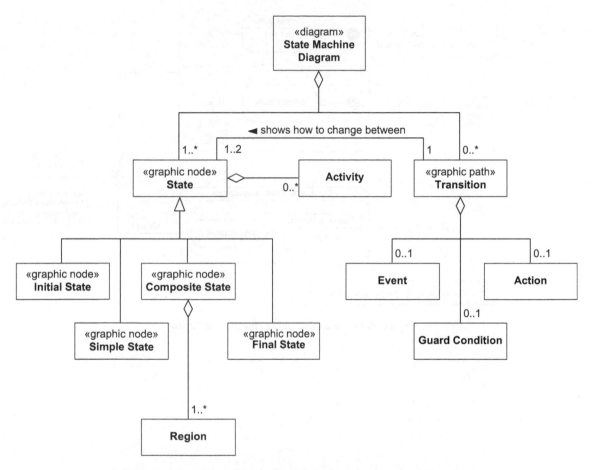

*Figure A.13   Partial meta-model for the* state machine diagram

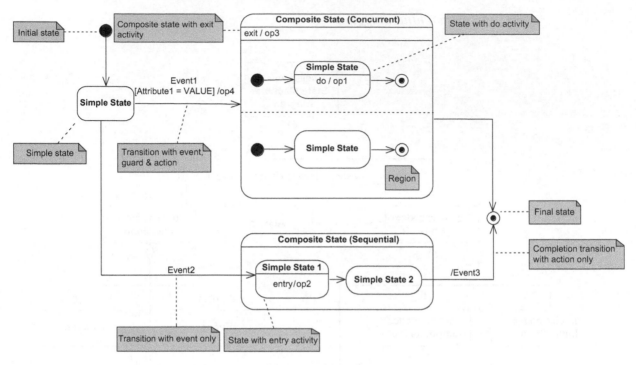

*Figure A.14*   State machine diagram *notation*

## *A.3.2    Sequence diagrams*

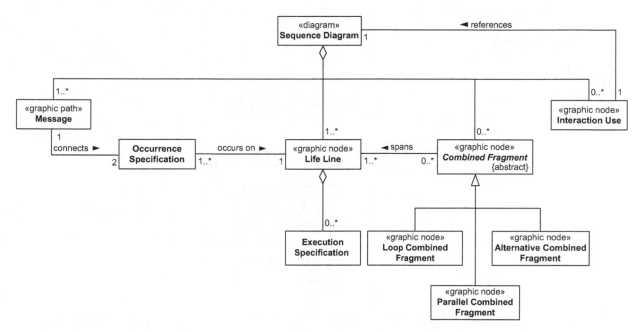

*Figure A.15    Partial meta-model for the* sequence diagram

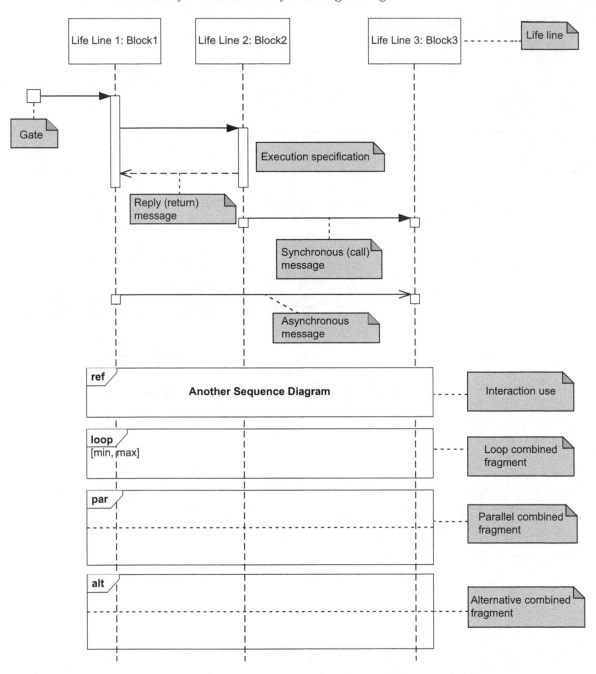

*Figure A.16    Sequence diagram* notation

## A.3.3    *Activity diagrams*

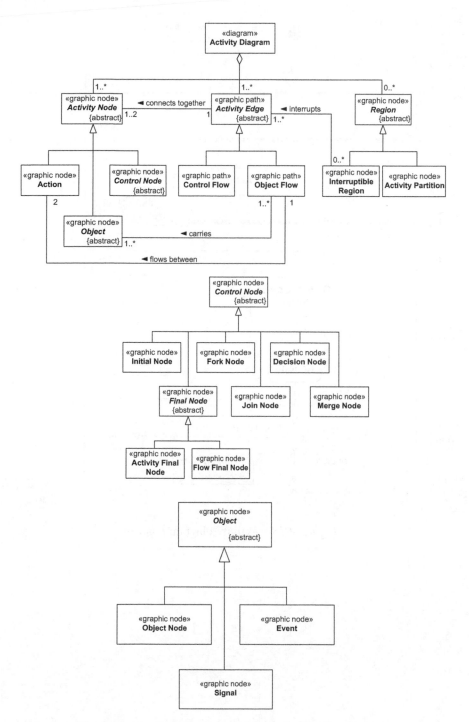

*Figure A.17    Partial meta-model for the* activity diagram

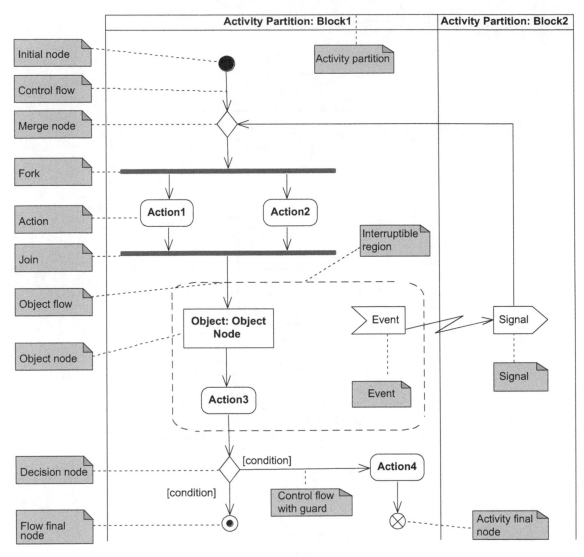

*Figure A.18*   Activity diagram *notation*

*A.3.4   Use case diagrams*

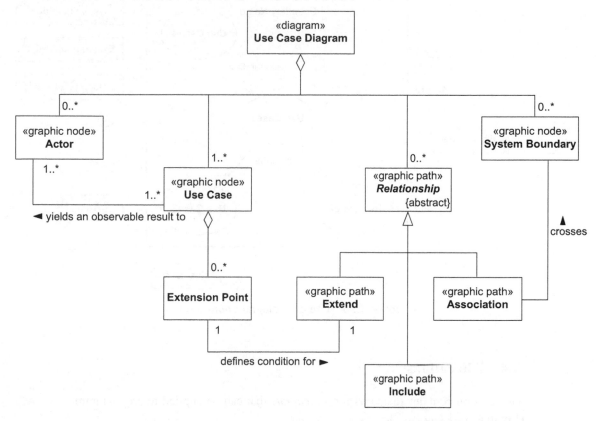

*Figure A.19   Partial meta-model for the* use case diagram

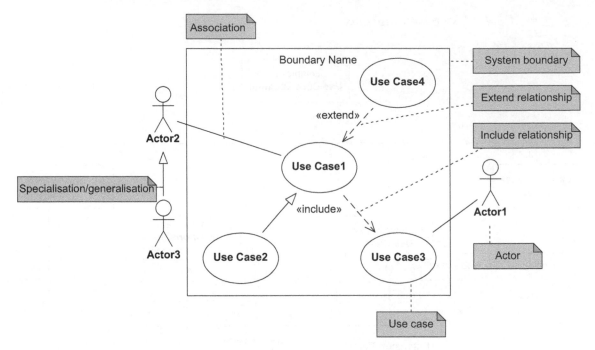

*Figure A.20*   Use case diagram *notation*

## A.4   Allocations

This section contains diagrams for *allocations* that can be applied to any diagram (Figures A.21 and A.22).

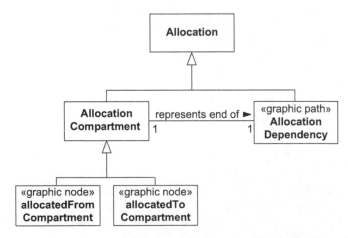

*Figure A.21*   *Partial meta-model for* allocations

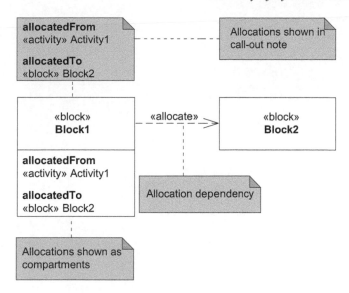

*Figure A.22*    Allocation *notation on* block definition diagram

# Reference

Holt J., Perry S.A. *SysML for Systems Engineering – 2nd Edition: A Model-based Approach*. Stevenage, UK: IET; 2013

# Summary of the Framework for Architectural Frameworks

## B.1  Introduction

This appendix provides a summary of the Framework for Architectural Frameworks (FAF) that was introduced briefly in Chapter 3 and discussed in detail in Chapter 19. No explanation is given in this Appendix; for details see Chapters 3 and 19, as well as Chapter 11 of 'SysML for Systems Engineering' (Holt and Perry, 2013).

The FAF was developed to improve the definition of Architectural Frameworks (AFs). It is designed to force anyone defining an AF to consider the following six questions:

1.  What is the purpose of the AF?
2.  What domain concepts must the AF support?
3.  What viewpoints are required?
4.  What is the purpose of each viewpoint?
5.  What is the definition of each viewpoint in terms of the identified domain concepts?
6.  What rules constrain the use of the AF?

The FAF addresses these six questions through six Viewpoints, each of which addresses a particular question:

1.  AF Context Viewpoint (AFCVp)

    o  What is the purpose of the AF?
       - Defines the context for the AF
       - Represents the AF Concerns in context, establishing why the AF is needed

2.  Ontology Definition Viewpoint (ODVp)

    o  What domain concepts must the AF support?
       - Defines the Ontology for the AF
       - Derived from the AF Context Viewpoint & and defines concepts that can appear on a Viewpoint

3.  Viewpoint Relationships Viewpoint (VRVp)
    - o What viewpoints are required?
        - ■ Shows the relationships between the Viewpoints that make up an AF
        - ■ Groups them into Perspectives. It is derived from the Ontology Definition Viewpoint

4.  Viewpoint Context Viewpoint (VCVp)
    - o What is the purpose of each viewpoint?
        - ■ Defines the Context for a particular Viewpoint
        - ■ Represents the Viewpoint concerns in context for a particular Viewpoint, establishing why the Viewpoint is needed. It is derived from the AF Context Viewpoint

5.  Viewpoint Definition Viewpoint (VDVp)
    - o What is the definition of each viewpoint in terms of the identified domain concepts?
        - ■ Defines a particular Viewpoint
        - ■ Shows the Viewpoint Elements (and hence the Ontology Elements) that appear on the Viewpoint

6.  Rules Definition Viewpoint (RDVp)
    - o What rules constrain the use of the AF?
        - ■ Defines the various Rules that constrain the AF.

The FAF is itself defined using the FAF. The FAF Views that realise each of the FAF Viewpoints are given in the following sections (Figures B.1–B.18).

## B.2    Architectural Framework Context View (AFCV)

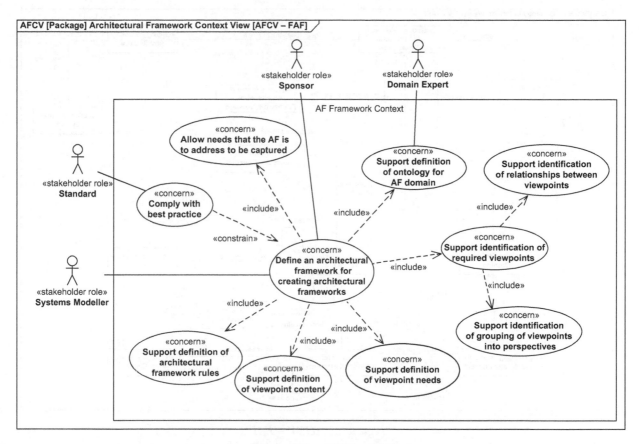

*Figure B.1    AFCV – FAF*

## B.3    Ontology Definition View (ODV)

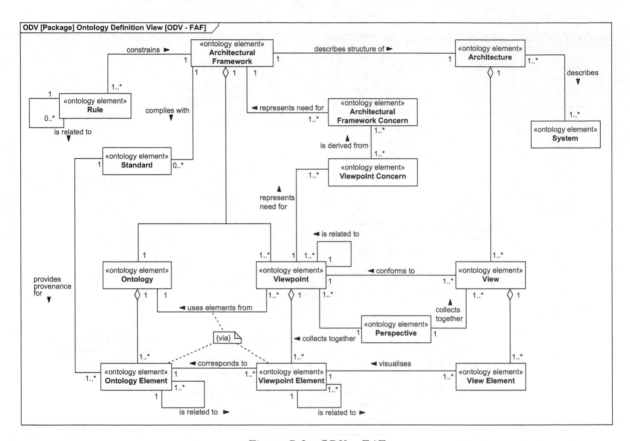

*Figure B.2    ODV – FAF*

## B.3.1   ODV – FAF – Expanding on Context

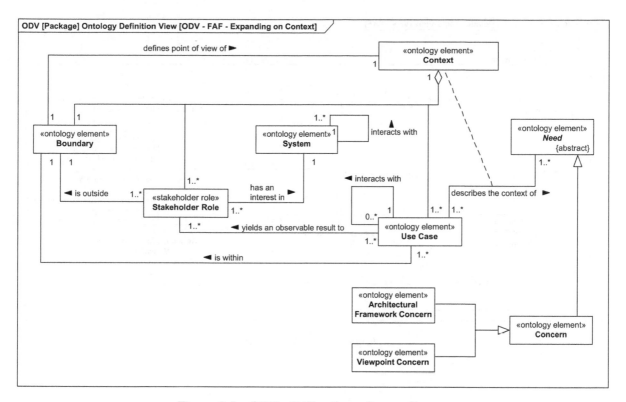

*Figure B.3    ODV – FAF – Expanding on Context*

## B.4   Viewpoint Relationships View (VRV)

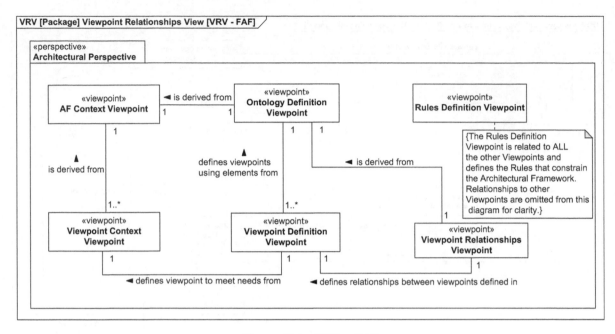

*Figure B.4    VRV – FAF*

## B.4.1 VRV – FAF – Complete Framework

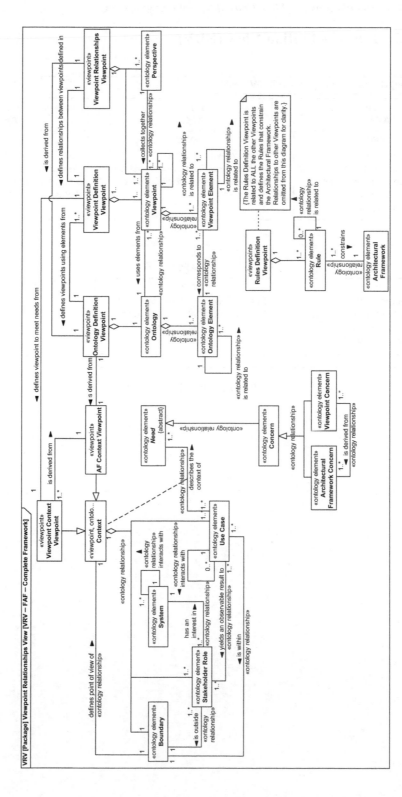

*Figure B.5  VRV – FAF – Complete Framework*

## B.5    Rules Definition View (RDV)

RDV [Package] Rules Definition View [RDV – FAF]

| «rule» **AF01** |
| --- |
| notes |
| *The definition of any Architectural Framework must include at least one instance (View) of each of the following Viewpoints: AFCV; ODV; VRV; VCV; VDV; RDV* |

| «rule» **AF02** |
| --- |
| notes |
| *Every Viewpoint in the Architectural Framework must be defined on a Viewpoint Definition Viewpoint.* |

| «rule» **AF03** |
| --- |
| notes |
| *Every Viewpoint Definition Viewpoint must be based on a corresponding Viewpoint Context Viewpoint.* |

| «rule» **AF04** |
| --- |
| notes |
| *Every Viewpoint in the Architectural Framework must appear on the viewpoint Relationships Viewpoint.* |

| «rule» **AF05** |
| --- |
| notes |
| *Every Viewpoint in the Architectural Framework must belong to one and only one Perspective.* |

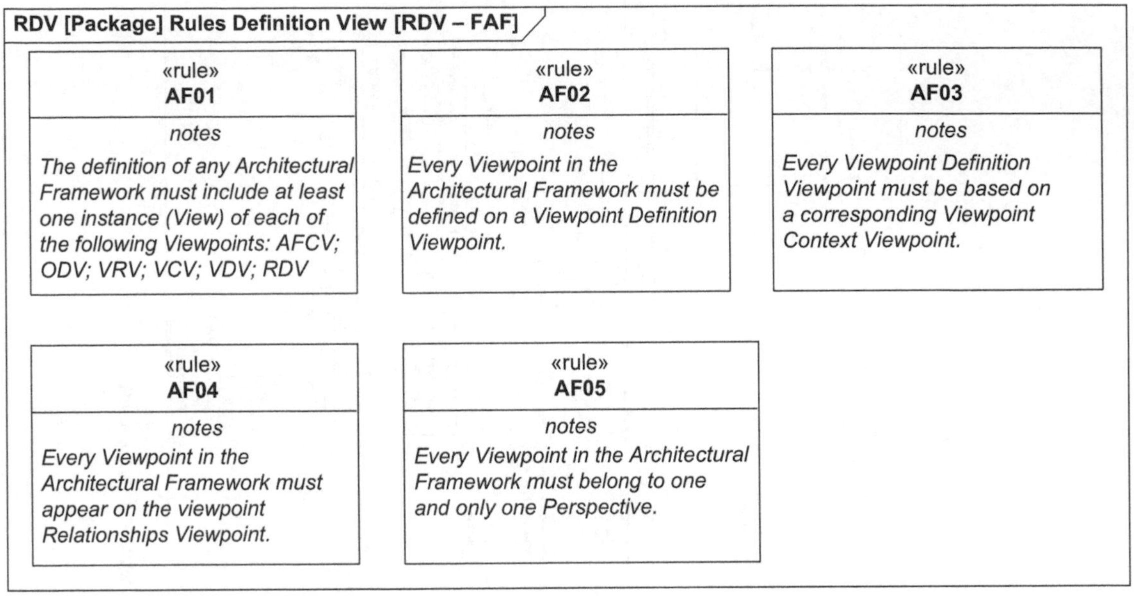

*Figure B.6    RDV – FAF*

## B.6   Viewpoint Context Views (VCVs)

### B.6.1   *VCV – AFCVp – FAF*

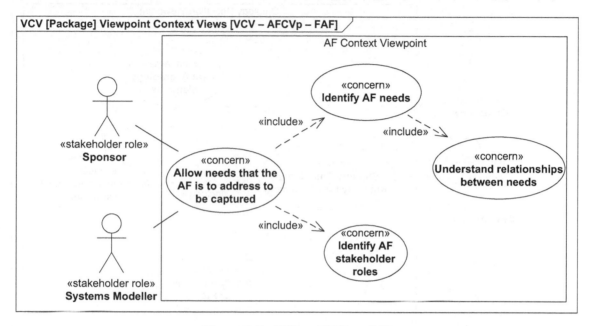

*Figure B.7    VCV – AFCVp – FAF*

## B.6.2   *VCV – ODVp – FAF*

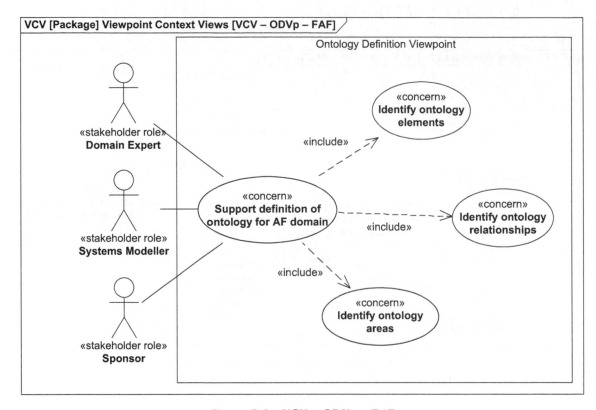

*Figure B.8    VCV – ODVp – FAF*

*B.6.3   VCV – VRVp – FAF*

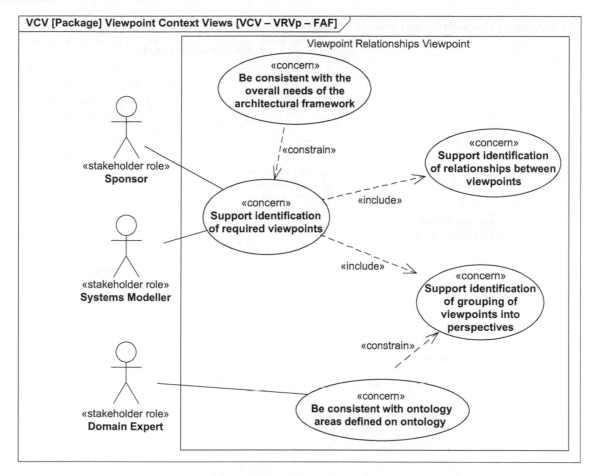

*Figure B.9   VCV – VRVp – FAF*

*B.6.4   VCV – RDVp – FAF*

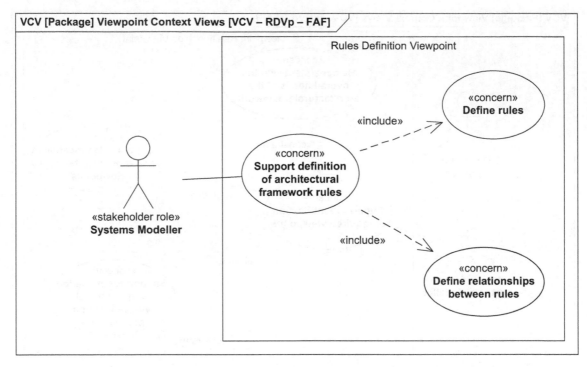

*Figure B.10   VCV – RDVp – FAF*

### B.6.5   VCV – VCVp – FAF

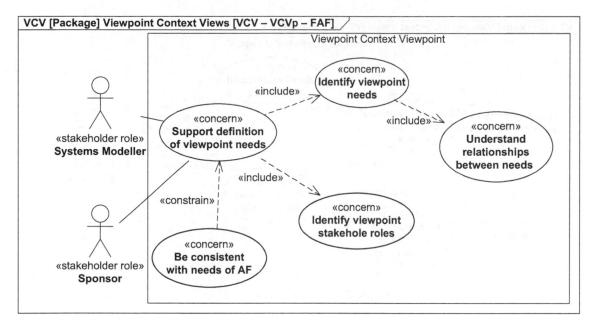

*Figure B.11    VCV – VCVp – FAF*

## B.6.6    VCV – VDVp – FAF

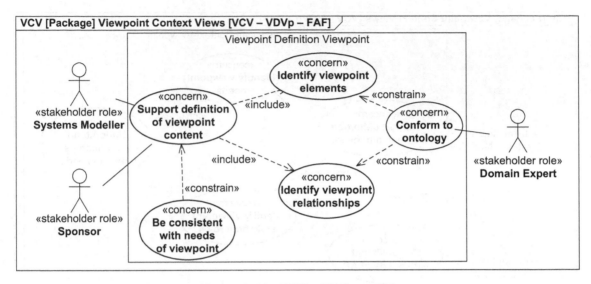

*Figure B.12    VCV – VDVp – FAF*

## B.7 Viewpoint Definition Views (VDVs)

### B.7.1 VDV – AFCVp – FAF

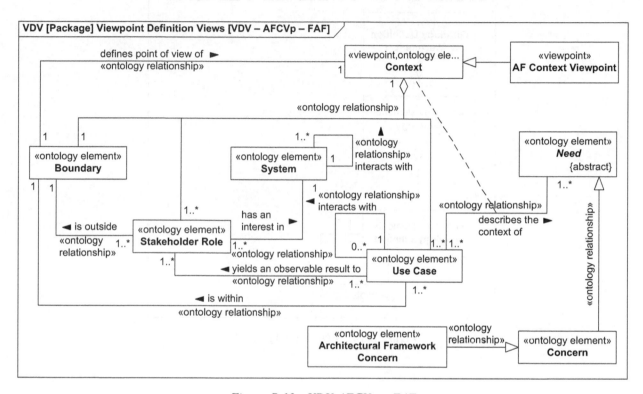

*Figure B.13 VDV AFCVp – FAF*

## B.7.2    *VDV – ODVp – FAF*

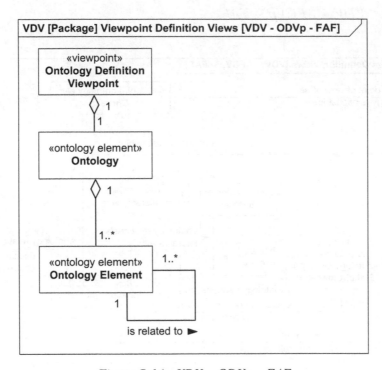

*Figure B.14    VDV – ODVp – FAF*

### B.7.3    VDV – VRVp – FAF

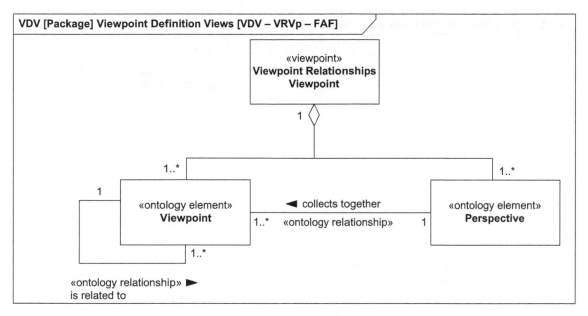

*Figure B.15    VDV – VRVp – FAF*

*B.7.4    VDV – RDVp – FAF*

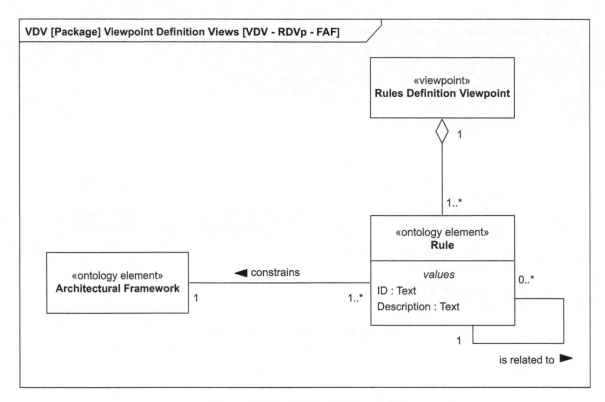

*Figure B.16    VDV – RDVp – FAF*

Summary of the Framework for Architectural Frameworks

### B.7.5 VDV – VCVp – FAF

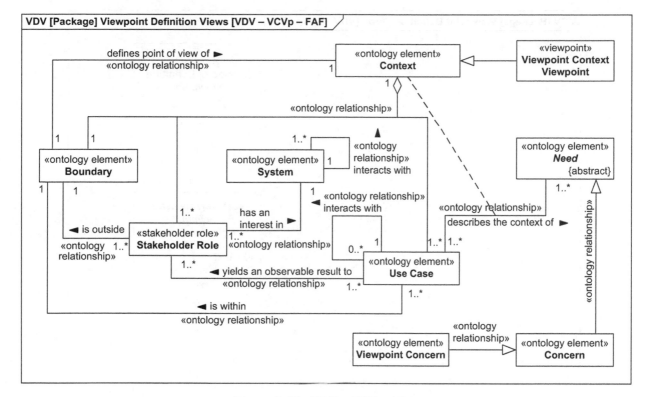

*Figure B.17 VDV – VCVp – FAF*

## B.8    VDV – VDVp – FAF

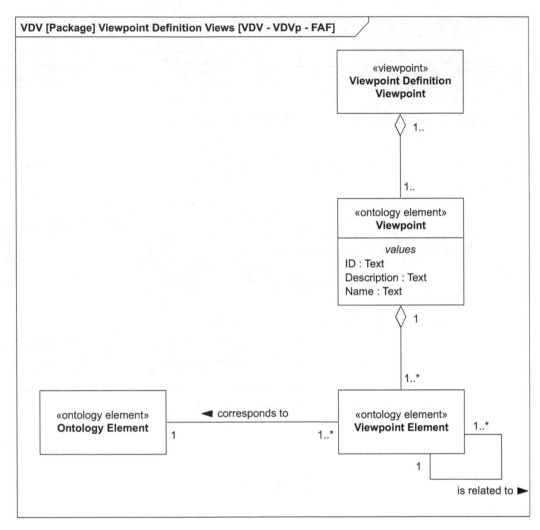

*Figure B.18    VDV – VDVp – FAF*

## Reference

Holt J. and Perry S. *SysML for Systems Engineering – 2nd Edition: A Model-Based Approach.* Stevenage, UK: IET; 2013.

*Appendix C*

# MBSE Ontology and glossary

## C.1 Introduction

This appendix provides a summary of the MBSE Ontology and the definitions of the concepts contained as defined in Holt and Perry (2013) to which the reader is referred for full details (Figure C.1).

## C.2 Ontology

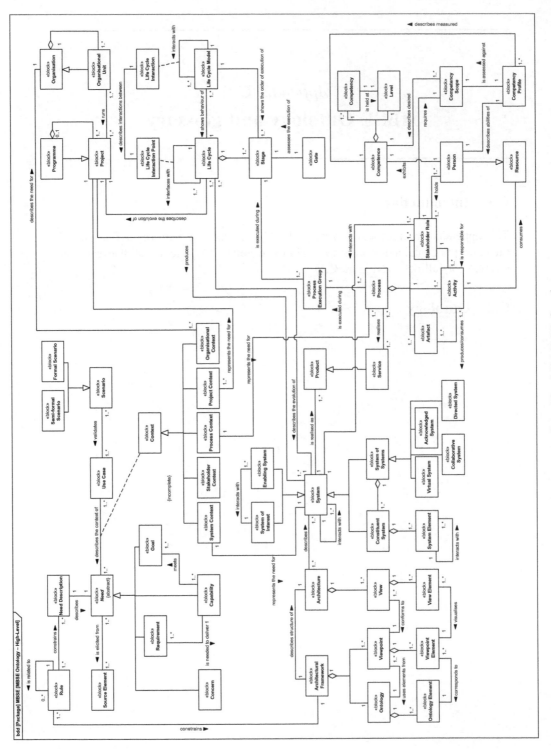

*Figure C.1   The full MBSE Ontology*

## C.3   Glossary

Acknowledged
System
A special type of System of Systems that has designated management and resources, and a consensus of purpose. Each Constituent System retains its own management and operation.

Activity
A set of actions that need to be performed in order to successfully execute a Process. Each Activity must have a responsible Stakeholder Role associated with it and utilises one or more Resource.

Architectural
Framework
A defined set of one or more Viewpoint and an Ontology. The Architectural Framework is used to structure an Architecture from the point of view of a specific industry, Stakeholder Role set, or Organisation. The Architectural Framework is defined so that it meets the Needs defined by one or more Architectural Framework Concern. An Architectural Framework is created so that it complies with zero or more Standard.

Architectural
Framework
Concern
Defines a Need that an Architectural Framework has to address.

Architecture
A description of a System, made up of one or more View. One or more related View can be collected together into a Perspective.

Artefact
Something that is produced or consumed by an Activity in a Process. Examples of an Artefact include: documentation, software, hardware and systems.

Capability
A special type of Need whose Context will typically represent one or more Project (as a Project Context) or one or more Organisational Unit (as an Organisational Context). A Capability will meet one or more Goal and will represent the ability of an Organisation or Organisational Unit.

Collaborative
System
A special type of System of Systems that lacks central management and resources, but has consensus of purpose.

Competence
The ability exhibited by a Person that is made up of a set of one or more individual Competency.

Competency
The representation of a single skill that contributes towards making up a Competence. Each Competency is held at a Level that describes the maturity of that Competency. There are four Levels defined for the MBSE Ontology.

Competency
Profile
A representation of the actual measured Competence of a Person and that is defined by one or more Competency. An individual's competence will usually be represented by one or more Competency Profile. A Competency Profile is the result of performing a competence assessment against a Competence Scope.

| | |
|---|---|
| Competency Scope | Representation of the desired Competence required for a specific Stakeholder Role and that is defined by one or more Competency. |
| Concern | A type of Need. Architectural Framework Concerns represent the Needs for an Architectural Framework; Viewpoint Concerns represent the Needs for a Viewpoint. |
| Constituent System | A special type of System whose elements are one or more System Element. |
| Context | A specific point of view based on, for example, Stakeholder Roles, System hierarchy level and, Life Cycle Stage. |
| Directed System | A special type of System of Systems that has designated management and resources, and a consensus of purpose. Each Constituent System retains its own operation but not management. |
| Enabling System | A special type of System that interacts with the System of Interest yet sits outside its boundary. |
| Formal Scenario | A Scenario that is mathematically provable using, for example, formal methods. |
| Gate | A mechanism for assessing the success or failure of the execution of a Stage. |
| Goal | A special type of Need whose Context will typically represent one or more Organisational Unit (as an Organisational Context). Each Goal will be met by one or more Capability. |
| Indicator | A feature of a Competency that describes knowledge, skill or attitude required to meet the Competency. It is the Indicator that is assessed as part of competency assessment. |
| Level | The level at which a Competency is held. |
| Life Cycle | A set of one or more Stage that can be used to describe the evolution of System, Project etc. over time. |
| Life Cycle Interaction | The point during a Life Cycle Model at which one or more Stage interact with each other. |
| Life Cycle Interaction Point | The point in a Life Cycle where one or more Life Cycle Interaction will occur. |
| Life Cycle Model | The execution of a set of one or more Stage that shows the behaviour of a Life Cycle. |
| Need | A generic abstract concept that, when put into a Context, represents something that is necessary or desirable for the subject of the Context. Often simply referred to as requirements. |
| Need Description | A tangible description of an abstract Need that is defined according to a pre-defined set of attributes. |
| Ontology | An element of an Architectural Framework that defines all the concepts and terms (one or more Ontology Element) that relate to any Architecture structured according to the Architectural Framework. |

| | |
|---|---|
| Ontology Element | The concepts that make up an Ontology. Each Ontology Element can be related to each other and is used in the definition of each Viewpoint (through the corresponding Viewpoint Element that makes up a Viewpoint). The provenance for each Ontology Element is provided by one or more Standard. |
| Organisation | A collection of one or more Organisational Units, it runs one or more Projects and will have its own Organisational Context. |
| Organisational Context | A Context defined from the point of view of a specific Organisation. |
| Organisational Unit | A special type of Organisation that itself can make up part of an Organisation. An Organisational Unit also runs one or more Projects and will have its own Organisational Context |
| Person | A special type of Resource, an individual human, who exhibits Competence that is represented by their Competency Profile. A Person also holds one or more Stakeholder Role. |
| Perspective | A collection of one or more View (and hence also one or more defining Viewpoint) that are related by their purpose. That is, one or more View which address the same architectural needs, rather than being related in some other way, such as by mode of visualisation, for example. |
| Process | A description of an approach that is defined by: one or more Activity, one or more Artefact and one or more Stakeholder Role. One or more Process also defines a Service. |
| Process Context | A Context defined from the point of view of a specific Process. |
| Process Execution Group | A set of one or more Processes executed in order for a specific purpose as part of a Stage. For example, a Process Execution Group may be defined based on a team, function, etc. |
| Product | Something that realises a System. Typical products may include, but are not limited to: software, hardware, Processes, data, humans, facilities, etc. |
| Programme | A special type of Project that is itself made up of one or more Project. |
| Project | One or more Project is run by an Organisational Unit in order to produce one or more System. |
| Project Context | A Context defined from the point of view of a specific Project. |
| Requirement | A property of a System that is either needed or wanted by a Stakeholder Role or other Context-defining element. Also, one or more Requirement is needed to deliver each Capability. |
| Resource | Anything that is used or consumed by an Activity within a Process. Examples of a Resource include money, locations, fuel, raw material, data and people. |

Rule                     A construct that constrains the attributes of a Need Description. A Rule may take several forms, such as equations, heuristics, reserved word lists, grammar restrictions. Or a construct that constrains an Architectural Framework (and hence the resulting Architecture) in some way, for example by defining one or more Viewpoint that are required as a minimum.

Scenario                 An ordered set of interactions between or more Stakeholder Role, System or System Element that represents a specific chain of events with a specific outcome. One or more Scenario validates each Use Case

Semi-formal              A Scenario that is demonstrable using, for example, visual
Scenario                 notations such as SysML, tables, text.

Service                  An intangible Product that realises a System. A Service is itself is realised by one or more Process.

Source Element           The ultimate origin of a Need that is elicited into one or more Need Description. A Source Element can be almost anything that inspires, effects or drives a Need, such as a Standard, a System, Project documentation, a phone call, an email, a letter, a book.

Stage                    A period within a Life Cycle that relates to its realisation through one or more Process Execution Group. The success of a Stage is assessed by a Gate.

Stakeholder              A Context defined from the point of view of a specific
Context                  Stakeholder Role.

Stakeholder Role         The role of anything that has an interest in a System. Examples of a Stakeholder Role include the roles of a Person, an Organisational Unit, a Project, a Source Element, and an Enabling System. Each Stakeholder Role requires its own Competency Scope and will be responsible for one or more Activity.

System                   A set of interacting elements organised to satisfy one or more System Context. Where the System is a System of Systems, then its elements will be one or more Constituent System, and where the System is a Constituent System then its elements are one or more System Element. A System can interact with one or more other System. The artefact being engineered that an Architecture describes.

System Context           A Context defined from the point of view of a specific System.

System Element           A basic part of a Constituent System.

System of Interest       A special type of System that describes the system being developed, enhanced, maintained or investigated.

System of Systems        A special type of System whose elements are one or more Constituent System and which delivers unique functionality not deliverable by any single Constituent System.

| | |
|---|---|
| Use Case | A Need that is considered in a specific Context and that is validated by one or more Scenario |
| View | The visualisation of part of the Architecture of a System, that conforms to the structure and content defined in a Viewpoint. A View is made up of one or more View Element. |
| View Element | The elements that make up a View. Each View Element visualises a Viewpoint Element that makes up the Viewpoint to which the View, on which the View Element appears, conforms. |
| Viewpoint | A definition of the structure and content of a View. The content and structure of a Viewpoint uses the concepts and terms from the Ontology via one or more Viewpoint Element that make up the Viewpoint. Each Viewpoint is defined so that it meets the needs defined by one or more Viewpoint Concern. |
| Viewpoint Concern | Defines a Need that a Viewpoint has to address. |
| Viewpoint Element | The elements that make up a Viewpoint. Each Viewpoint Element must correspond to an Ontology Element from the Ontology that is part of the Architectural Framework. |
| Virtual System | A special type of System of Systems that lacks central management and resources, and no consensus of purpose. |

# Reference

Holt, J. and Perry, S.A. *SysML for Systems Engineering – 2nd Edition: A Model-Based Approach*. Stevenage, UK: IET Publishing; 2013.

# The 'Pattern Definition and Realisation (PaDRe)' Processes

## D.1   Introduction

This appendix presents a set of Processes for Pattern definition. These Processes, known as PaDRe (from **Pa**ttern **D**efinition and **Re**alisation) are based on the ArchDeCon processes found in appendix F of Holt and Perry (2013). The processes are essentially the same as the ArchDeCon processes (used for the definition of Architectural Frameworks (AFs)) but renamed so that they refer to Patterns rather than AFs. The only exception to this is that the 'Pattern context view' produced by the Processes is an instance of an 'AF Context Viewpoint' rather than a 'Pattern Context Viewpoint'. This is because, as discussed in Chapter 3, the underlying Framework used for Pattern definition is one which was originally created for the definition of AFs. As noted in Chapter 3, feel free to rename the 'AF Context Viewpoint' to be the 'Pattern Context Viewpoint' if you wish.

## D.2   Requirements Context View (RCV)

The basic Needs for the PaDRe processes are shown in Figure D.1.

Figure D.1 shows the Context for the definition of a set of processes for the definition of Patterns. The main Use Case that must be fulfilled is to 'Define processes for creating patterns', constrained by 'Comply with best practice'. In order to 'Define processes for creating patterns' it is necessary to:

- 'Allow needs that the pattern is to address to be captured' – When defining a Pattern, it is important that the Needs that the Pattern is to address can be captured, in order to ensure that the Pattern is fit for purpose.
- 'Support definition of ontology for pattern domain' – When defining a Pattern, it is essential that the concepts, and the relationships between them, are defined for the domain in which the Pattern is to be used. This is the Ontology that forms the foundational basis of the definition of the Pattern's Viewpoints. Such an Ontology ensures the consistency of the Pattern. The Processes must support such a definition of an Ontology.

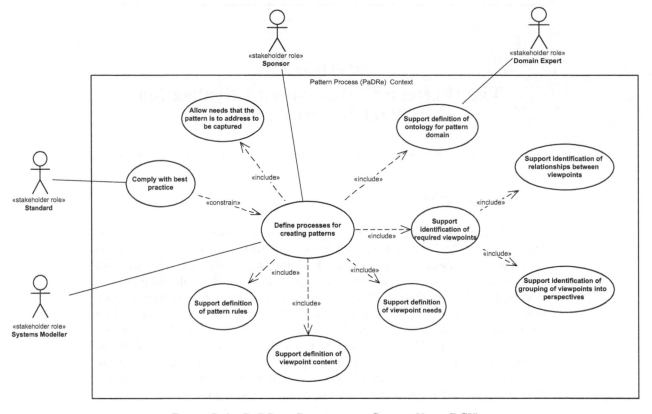

*Figure D.1    PaDRe – Requirement Context View (RCV)*

- 'Support definition of pattern rules' – Often, when defining a Pattern, it is often necessary to constrain aspects of the Pattern through the definition of a number of constraining Rules. It is therefore essential that the Processes for the definition of a Pattern support the definition of such Rules.

- 'Support definition of viewpoint content' – A Pattern is essentially a number of Viewpoints that conform to an Ontology. Therefore, when defining a Pattern, it is essential that each Viewpoint can be defined in a consistent fashion that ensures its conformance to the Ontology.

- 'Support definition of viewpoint needs' – In order to define the Viewpoints that make up a Pattern, it is essential that the Needs of each Viewpoint be clearly understood in order to ensure each Viewpoint is fit for purpose and that the Viewpoints defined meet the overall Needs for the Pattern.

- 'Support identification of required viewpoints' – The Viewpoints that make up the Pattern need to be identified. As well as supporting such an identification, the Processes must also 'Support identification of relationships between viewpoints' and 'Support identification of grouping of viewpoints into perspectives'.

## D.3    Process Content View (PCV)

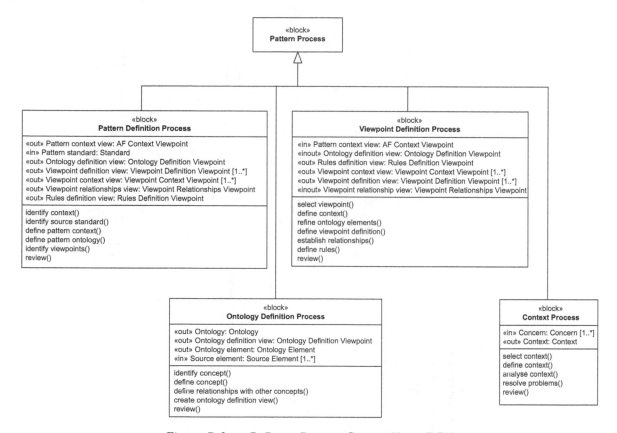

*Figure D.2    – PaDre – Process Content View (PCV)*

The PaDre Processes shown in Figure D.2 may be summarised as:

- 'Pattern Definition Process' – The aim of this Process is to understand the underlying need for the Pattern. It uses the other three PaDre Processes to: identify Concerns for the Pattern and put them into Context, defines an Ontology, defines Viewpoints, Perspectives and Rules.
- 'Ontology Definition Process' – The aim of this Process is to identify and define the main concepts and terms used for the Pattern in the form of an Ontology.
- 'Viewpoint Definition Process' – The aim of this Process is to identify and define the key Viewpoints and to classify them into Perspectives. It also defines any Rules that constrain the Viewpoints and Pattern.
- 'Context Process' – The aim of this Process is to create a Context that can be used to create either an 'Pattern context view' or a 'Viewpoint context view'. The remaining "seven views" process modelling Views are shown in Figures D.3–D.11.

**D.4 Stakeholder View (SV)**

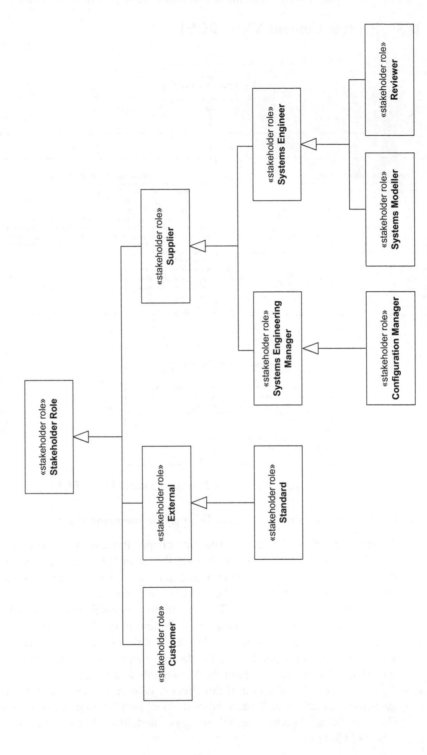

Figure D.3   PaDre – Stakeholder View (SV)

## D.5   Information View (IV)

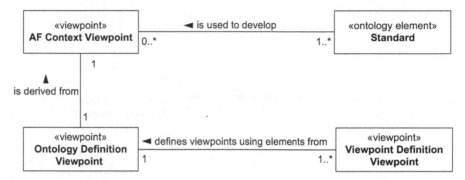

Figure D.4   *PaDre – Information View (IV) for Pattern Definition Process Artefacts*

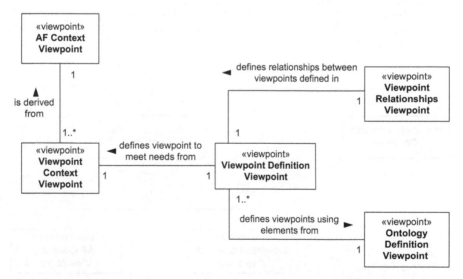

Figure D.5   *PaDre – Information View (IV) for Viewpoint Definition Process Artefacts*

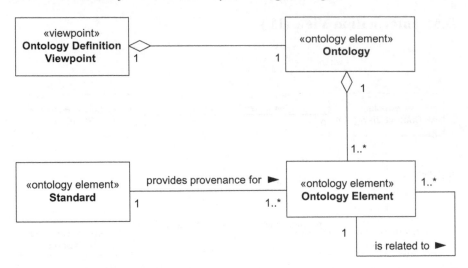

*Figure D.6    PaDre – Information View (IV) for Ontology Definition Process Artefacts*

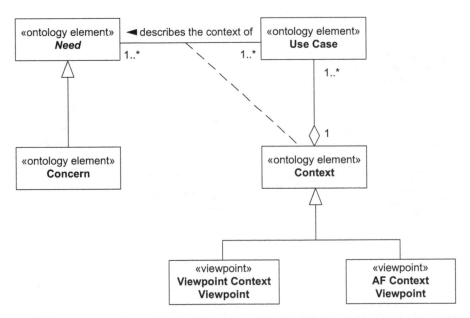

*Figure D.7    PaDre – Information View (IV) for Context Process Artefacts*

## D.6   Process Behaviour View (PBV)

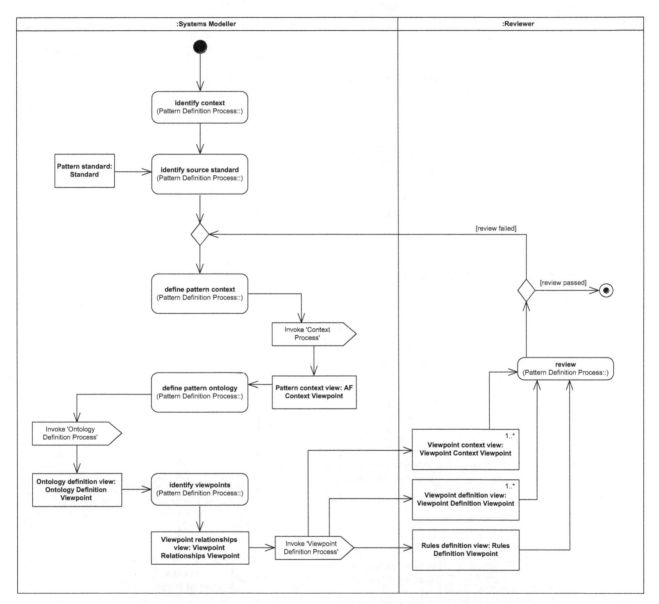

*Figure D.8   PaDre – Process Behaviour View (PBV) for Pattern Definition Process*

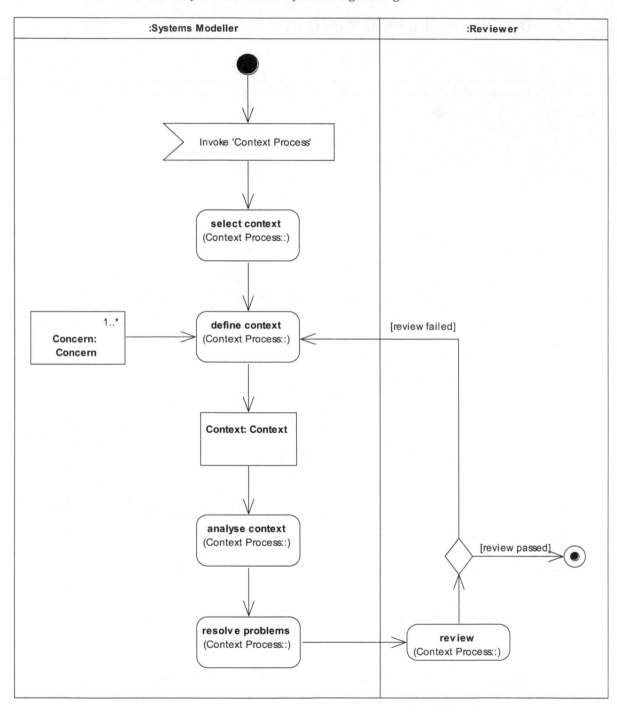

*Figure D.9    PaDre – Process Behaviour View (PBV) for Context Process*

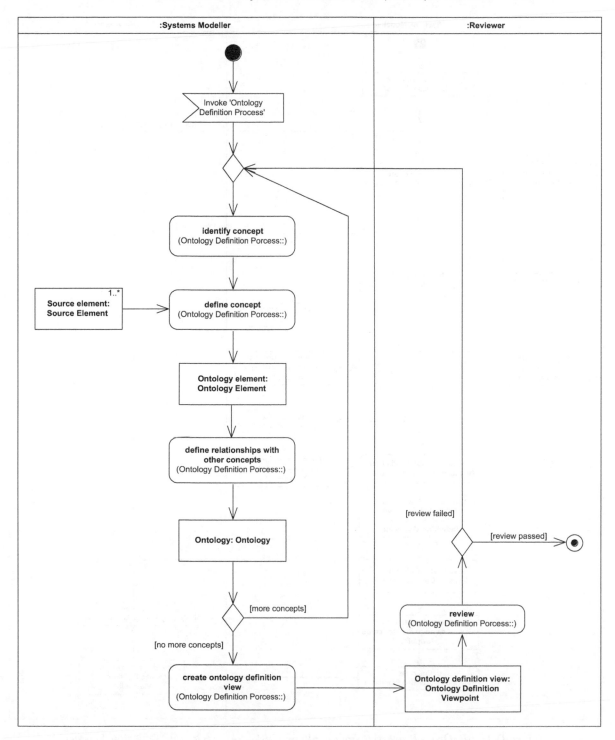

*Figure D.10   PaDre – Process Behaviour View (PBV) for Ontology Definition Process*

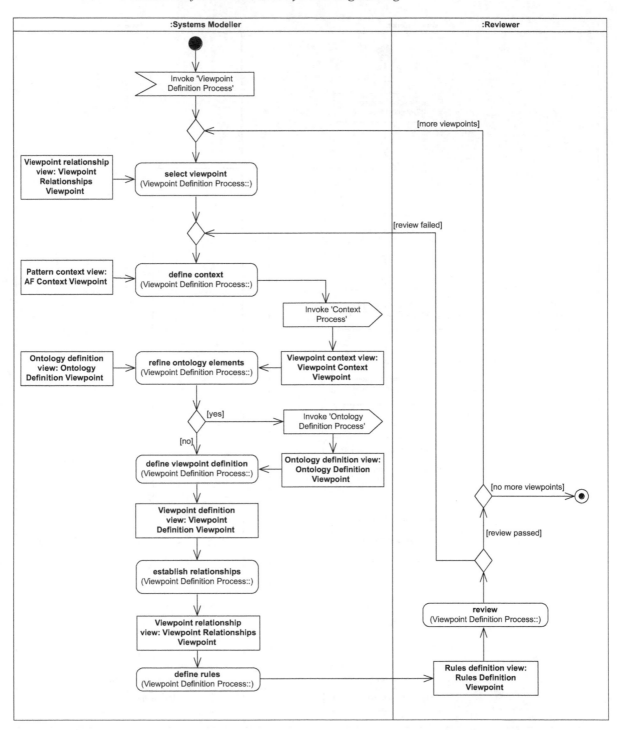

*Figure D.11    PaDre – Process Behaviour View (PBV) for Viewpoint Definition Process*

## Reference

Holt J. and Perry S. *SysML for Systems Engineering – 2nd Edition: A Model-Based Approach*. Stevenage, UK: IET Publishing; 2013.

# Index